AF361860

# Bioenergy and Biochar: Repurposing Waste to Sustainable Energy and Materials

# Bioenergy and Biochar: Repurposing Waste to Sustainable Energy and Materials

Editors

**Dimitrios Kalderis**
**Vasiliki Skoulou**

MDPI • Basel • Beijing • Wuhan • Barcelona • Belgrade • Manchester • Tokyo • Cluj • Tianjin

*Editors*
Dimitrios Kalderis
Department of Electronics Engineering
Hellenic Mediterranean University
Chania
Greece

Vasiliki Skoulou
Biochemistry and Chemical Engineering Department
University of Hull
Hull
United Kingdom

*Editorial Office*
MDPI
St. Alban-Anlage 66
4052 Basel, Switzerland

This is a reprint of articles from the Special Issue published online in the open access journal *Energies* (ISSN 1996-1073) (available at: www.mdpi.com/journal/energies/special_issues/ Bioenergy_Biochar).

For citation purposes, cite each article independently as indicated on the article page online and as indicated below:

LastName, A.A.; LastName, B.B.; LastName, C.C. Article Title. *Journal Name* **Year**, *Volume Number*, Page Range.

**ISBN 978-3-0365-1856-5 (Hbk)**
**ISBN 978-3-0365-1855-8 (PDF)**

# Contents

# About the Editors

**Dimitrios Kalderis** Dr. Dimitrios Kalderis is Assoc. Professor in the Department of Electronics Engineering, at the Hellenic Mediterranean University, Chania, Greece. He obtained a B.Sc. in Chemistry from the School of Chemistry, University of Leeds, UK, and a Ph.D. in Environmental Chemistry from the same university. His research interests involve the processing of biomass/agricultural waste for the production of added-value materials such as biochars and hydrochars, the hydrothermal carbonization of industrial wastewaters and solid waste, and the remediation of soils contaminated with organic substances. He has participated in more than 10 EU- and nationally funded projects and he has published more than 50 papers in peer-reviewed international journals.

**Vasiliki Skoulou** Dr. Vasiliki Skoulou is Assoc. Professor (Senior Lecturer) and Director of Research of Chemistry-Biochemistry and Chemical Engineering Departments at the University of Hull, UK. She holds a B.Sc., Integrated Masters, and a Ph.D. in Chemical Engineering from the Aristotle University of Thessaloniki, Greece, and an M.Sc. in Environmental Protection and Sustainable Development. Her research focuses on biomass waste pretreatments, low carbon thermal/thermochemical treatments (mainly pyrolysis and gasification) of lignocellulosic biomass residues, solid fuels, waste, and their blends, as well as designing of thermochemical reactors for waste to energy, $H_2$, and char-activated carbons production. She has participated in more than 25 research projects in the field of low-carbon biomass waste to energy, and the outcomes of her research are included in more than 75 publications in book chapters, journals, and conference proceedings.

# Preface to "Bioenergy and Biochar: Repurposing Waste to Sustainable Energy and Materials"

The following summary of the Special Issue papers was kindly prepared by Thomas R. Miles, Executive Director of the United States Biochar Initiative. This Special Issue on "Bioenergy and Biochar: Repurposing Waste to Sustainable Energy and Materials"comprises 11 papers that explore creative pathways to zero waste, greenhouse gas (GHG) reduction, and circular economies through recycling of nutrients, feedstock production on marginal land and natural grasslands, and conversion of agricultural and wood residues. Renewable, low-carbon products include coffee cup lids made from starch-based thermoplastic and coffee waste biochar, fuels and biochars from coconut husks and shells, biochar-amended manure pellets for rice cultivation, hydrochars from sewage sludge and food waste, ethanol from sugarcane grown on marginal land, mineral soil amendments from biomass boiler ash, lipids for biofuels from meat wastes, low greenhouse gas (GHG) ethanol from wood, fermentable sugars from semi-natural grasslands, and methane from microalgae-enhanced anaerobic digestion.

Diaz et al. produced biodegradable containers that can be degraded through processes such as composting or bio-digestion at the end of life to demonstrate closed-loop systems for organic waste. They used thermoplastic starch to replace traditional plastic and thermoformed it with polycaprolactone. They then used coffee waste biochars as fillers. The properties of the materials were tested to show that the coffee waste biochars could be reused. They conclude that starch and biochar can be used for manufacturing thermoformed containers. This is a continuation of ongoing research.

Two papers evaluate the suitability of coconut wastes as fuels and biochar. Obeng et al. in Ghana gasified coconut husks and shells to make green charcoal. They sun-dried coconut residues, which constitute 62–65% of the whole coconut fruit, and gasified them in a simple top-lit updraft gasifier, or TLUD. TLUDs are widely used to produce biochars for soil but have not been used on coconut shells and husks. Heating values of the char increased by 42% compared with the uncharred wastes. Emissions from the TLUD exceeded WHO standards but can be optimized through design. The authors recommend a switch from open burning to carbonization in a controlled system and briquetting to maximize calorific value and minimize smoke emissions in domestic cooking.

Coconut waste is abundant in 90 countries. In spite of the extensive production of biochar from coconut shell for charcoal and activated carbon, there is limited literature on coconut shell pyrolysis. Sarkar and Wang obtained coconut shells from Bangladesh and conducted detailed studies to determine the product yields and characteristics at increasing pyrolysis temperatures. The authors found that increasing temperature between 400 and 600 C resulted in important changes in yield and physical and chemical characteristics of the char, oil, and gas. This will be useful for those wanting to promote solid, liquid, or gaseous products from waste coconut shell.

Methods to use biochars to recycle manures and mitigate greenhouse gas emissions in rice cultivation were tested by Shin et al. in Korea. They tested pelletized biochar, manure, and animal waste compost amended fertilizers as environmentally safe application methods to mitigate non-source pollution and to reduce nutrient loss from drift and surface losses. Supplemented biochar manure pellets (SBMP) were made from 40% rice husk biochar combined with 60% composted pig manure. Urea, phosphate, and potassium chloride were added in various combinations and applied in a neutral clay loam soil. Paddy water quality showed that the SBMP can mitigate the loss of

"

nitrogen and phosphorous. Silicon increased and nutrient release was slower. Carbon sequestration was measured, and the cost targets for GHG reduction were established. Authors conclude that the application of SBMP fertilizers can contribute to reducing the agro-environmental impacts of runoff and enhance sequestration and rice yield.

Vardiambasis et al. analyzed existing research on hydrothermal carbonization (HTC) using advanced techniques to determine the focus of research and to determine correlations between feedstock and product qualities. They reviewed publications between 2014 and 2020, which indicated sewage sludge and food waste to be the most popular feedstocks for HTC. They conducted a statistical analysis of the key properties to establish correlations to guide analysis. They developed a series of modules using artificial neural networks (ANN) and used the models to predict higher heating values (HHVs) from carbon and other fuel elements. The work is a fascinating review of HTC research. It demonstrates a series of useful tools for literature review with useful outputs.

Ash and fly ash were traditionally applied on agricultural or forest lands as a mineral ash supplement. Today, a large proportion of fly ash from biomass plants is landfilled. Land application in Europe has been restricted by regulations that are based on contaminants in coal fly ash. Bubbling and circulating fluidized bed boilers have become the predominant biomass technologies in Europe. Jarosz-Krzeminska and Poluszynska show through extensive analysis that feedstock supplies and combustion technologies have improved biomass fly ash. They examine the physical and chemical properties of the fly ash, micro- and macronutrients, contaminants, non-essential elements, and the bioavailability of elements. They investigated the speciation of metals and acute toxicity of fly ash amendments to plant germination and growth. They show that fly ash from bubbling and circulating fluidized bed boilers have different characteristics due to process conditions and feedstocks. Circulating fluidized beds (CFB) operate at higher temperatures and recirculate the fly ash. Bubbling fluidized bed (BFB) fly ash was richer in potassium, phosphorous, carbonates, and micronutrients than fly ash from CFBs. The BFBs also have fewer contaminants. They attribute some differences to mixtures of feedstocks, with the BFBs firing higher percentages (20%) of agricultural residues including straws and sunflower husks. They did not find toxic effects on plant growth or germination from either technology, so they concluded that biomass fly ash should be used as soil amendments instead of landfilling.

Li et al. in China explore the potential conversion of meat wastes to biodiesel through the production of microbial lipids using strains of yeast. The team used amino acid (AA) blends, which represented sheep viscera and fish waste as carbon sources for lipid production with the oleaginous yeast *Rhodosporidium toruloides*. The lipid products and fatty acids compared favorably with those produced from vegetable oils from maize stover and palm. They concluded that further research is needed to identify cost-effective protein wastes, more robust oleaginous yeast strains, and advanced bioprocesses.

The demand for ethanol in the European Union for blending with petrol and diesel is expected to double to achieve 14% by 2030. Incentives are provided for processes that produce ethanol with lower GHG emissions. Sawdust from wood processing is an abundant lignocellulosic feedstock in Sweden. Haus et al. evaluate the economic competitiveness of lignocellulosic ethanol compared with agricultural-based ethanol fuels, which are imported. They found that the savings in GHG emission from the sawdust-based ethanol was 93% compared with 68% for the mainly crop-based ethanol, which could result in a 40% increase in price for sawdust-based ethanol. The authors modeled a 200,000 dry mt per year plant to see if the increased economic advantage was sufficient to promote large-scale commercial production. Various alternatives for energy recovery and feedstock procurement were analyzed. The authors determined that lignocellulosic ethanol could be viable and that the incentives could be useful in the long term but that they were insufficient to offset the high short-term risks of large-scale production.

Unmanaged, semi-natural grasslands are a potential biomass resource in Europe, but diverse species and highly variable factors challenge the conversion of these herbaceous species to fermentable sugars. Mezule et al. evaluate potential fermentable sugar yields and overall productivity from various grasslands habitats, which are common in a temperate climate and classified under European Union habitat codes. They used non-commercial enzymes from white-rot fungi as agents in hydrolysis. They evaluated habitat type, seasonality, cutting time, weather, and solids content in the biomass and other factors in two municipalities in Latvia. Of six habitat types, the highest yields of fermentable sugars were obtained from lowland grasslands and scrubland facies on calcareous substrates. The highest average yields were from lowland meadows and the lowest were from xeric sand calcareous grasslands. These correspond to yields from semi-natural grasslands in Estonia, central Germany, and Denmark. Average dry matter yields ranged from 1 to 6 tons per ha. The reasons for variations in yield are discussed. Additional research is needed to determine other factors that could impact production on these grasslands. Authors conclude that fermentable carbohydrate production can be used as an alternate strategy to grazing.

To address China's needs for additional sources of renewable fuels, Peng et al. simulate the production of sugarcane on marginal and cultivated land in the Southern province of Guangxi. They located potential lands through statistical methods while avoiding lands reserved for other uses. They then used a modification of the APSIM sugarcane model to simulate the growth in the selected areas. They verified model results through field testing and GIS techniques. The results allowed them to estimate the potential ethanol production, which resulted in opportunities to export to other provinces. They point out that additional study is needed to ensure that the lands they have identified are not subject to environmental hazards not considered in the simulation or GIS data.

Anaerobic digestion is an important conversion pathway for electricity, heat, and transportation in Europe. Debowki et al. investigate the effect of adding microalgae to common feedstocks such as cattle manure and maize silage for biogas production. Algal biomass is a source of nitrogen and microelements for the growth of microorganisms. Microalgae have a high growth rate and do not compete with crops for feed or food. They have a high photosynthetic efficiency, fast growth rate, potential to utilize $CO_2$, and resistance to contamination and can be cultured in areas not suitable for other uses. Microalgae culture was raised in photobioreactors and mixed with cattle slurry and maize silage. Six different species were tested. Adding microalgae improved biogas yield and composition. Methane increased. They found the highest methane production when the ratio of microalgae to feedstock was added at 20–40% v/v. There was no change in efficiency or other parameters. They found a strong correlation between methane production and C/N ratios, Anaerobic digestion with microalgae was limited by high protein and low C/N ratios which can be aided by co-digestion with carbon-rich feedstocks.

**Dimitrios Kalderis, Vasiliki Skoulou**

*Editors*

*Article*

# Thermoformed Containers Based on Starch and Starch/Coffee Waste Biochar Composites

**Carlos A. Diaz** [1,2,*], **Rahul Ketan Shah** [1], **Tyler Evans** [1], **Thomas A. Trabold** [2,3]
**and Kathleen Draper** [2,4]

[1]   Department of Packaging Science, Rochester Institute of Technology, Rochester, NY 14623, USA;
      rks3915@rit.edu (R.K.S.); tje2753@rit.edu (T.E.)
[2]   Cinterest LLC, Rochester, NY 14623, USA; tatasp@rit.edu (T.A.T.); biocharro2@gmail.com (K.D.)
[3]   Department of Sustainability, Golisano Institute for Sustainability, Rochester Institute of Technology,
      Rochester, NY 14623, USA
[4]   Ithaka Institute for Carbon Intelligence, Rochester, NY 14424, USA
*    Correspondence: cdamet@rit.edu

Received: 6 October 2020; Accepted: 14 November 2020; Published: 19 November 2020

**Abstract:** Biodegradable containers support zero-waste initiatives when alternative end-of-life scenarios are available (e.g., composting, bio digestion). Thermoplastic starch (TPS) has emerged as a readily biodegradable and inexpensive biomaterial that can replace traditional plastics in applications such as food service ware and packaging. This study has two aims. First, demonstrate the thermoformability of starch/polycaprolactone (PCL) as a thermoplastic material with varying starch loadings. Second, incorporate biochar as a sustainable filler that can potentially lower the cost and enhance compostability. Biochar is a stable form of carbon produced by thermochemical conversion of organic biomass, such as food waste, and its incorporation into consumer products could promote a circular economy. Thermoformed samples were successfully made with starch contents from 40 to 60 wt.% without biochar. Increasing the amount of starch increased the viscosity of the material, which in turn affected the compression molding (sheet manufacturing) and thermoforming conditions. PCL content reduced the extent of biodegradation in soil burial experiments and increased the strength and elongation at break of the material. A blend of 50:50 starch:PCL was selected for incorporating biochar. Thermoformed containers were manufactured with 10, 20, and 30 wt.% biochar derived from waste coffee grounds. The addition of biochar decreased the elongation at break but did not significantly affect the modulus of elasticity or tensile strength. The results demonstrate the feasibility of using starch and biochar for the manufacturing of thermoformed containers.

**Keywords:** starch; biochar; coffee waste; polycaprolactone; bioplastics; biodegradation

---

## 1. Introduction

Zero-waste initiatives call for waste to be either recyclable or compostable. Some municipalities in the United States (US) have programs to voluntarily separate organic waste, which is collected and subsequently composted or processed in an anaerobic digester. In this scenario, packaging and single-use items that are readily degradable present an opportunity to support and enhance closed-loop systems for organic waste.

Thermoplastic starch (TPS) has emerged as a readily biodegradable and inexpensive biomaterial that can replace traditional plastics in applications such as food service and packaging [1]. Our previous study [2] investigated the mechanical performance of TPS blends and polycaprolactone (PCL). A brittle–ductile transition was observed with the addition of PCL, and the degree of anaerobic biodegradation correlated with the amount of TPS. However, the preparation of TPS using water and glycerol showed inconsistencies from batch to batch, and it was susceptible to aging [2–4]. Therefore, the development of TPS-based products would benefit from a manufacturing process that avoids the use of water or glycerol.

Here, a direct mixing of starch and PCL is proposed, which bypasses some of the drawbacks outlined with TPS and could facilitate scale-up production. Additionally, the manufacturing of composites using biochar is presented as a means to enhance compostability and valorize a byproduct from the conversion of organic waste, thus promoting a circular economy [5]. Biochar is produced by pyrolysis of organic matter at high temperatures under zero-oxygen conditions [6]. This technique creates a highly stable carbon-rich material with physical properties, such as density, surface area, and porosity, that can be controlled by selecting critical process parameters, including heating rate, maximum temperature (typically in the range of 400 to 800 °C), and residence time [7]. Biochar has been highlighted in the Intergovernmental Panel on Climate Change (IPCC) Special Report: Global Warming of 1.5 °C as one of the carbon dioxide removal technologies that can help mitigate climate change. In the process of gasification, some oxygen is introduced to the system (well below the stoichiometric requirement for full combustion), and this may improve biochar quality in some cases, but at the cost of lower yield [8].

The research reported in this paper evolved from our prior work in developing bioplastic–biochar composite packaging that offers improved end-of-life management options while enabling valorization of food waste that would otherwise be landfilled. This work builds upon a rapidly expanding collection of studies published since 2015, summarized in Table 1, that have documented the potential advantages of using biochar as an additive in plastic products due to its favorable characteristics, including high surface area and long-term chemical and physical stability [9–27]. Reported improvements in the performance of polymer–biochar composites include enhanced water adsorption, thermal resistance, and stiffness. The added benefits of eliminating the disposal of organic wastes in landfills (potentially generating methane emissions) and sequestering carbon in the biochar material itself further contribute to its suitability for integration into circular manufacturing systems. In selecting a feedstock suitable for biochar production, it is desirable to identify a waste stream that is generally homogeneous, available in large quantities at low or zero cost, with minimal temporal and/or geographic variations. Waste coffee grounds were determined to satisfy all these requirements and were thus utilized in developing the prototype composite containers described below. It should be noted that there has been significant prior work reported on the use of coffee waste in sustainable material development, both in its raw state (e.g., [28–31]) and after thermochemical conversion to biochar [13,16,21,24,32]. Our results extend this earlier research by improving the understanding of bioplastic–coffee waste biochar composites that can meet the required functional specifications while enhancing degradability at the end of life. In addition, based on our prior research and the literature cited above, biochar has the potential to reduce the cost of thermoplastic materials by using waste feedstocks to displace common fillers and colorants, such as carbon black.

Abdelwahab and coworkers [15] investigated the use of biochar on injection-molded polypropylene and compared it to glass fiber and talc. Compared to propylene alone and the other fillers, biochar showed better thermal stability as measured by the coefficient of linear thermal expansion. Arrigo and coworkers [24] incorporated biochar from spent coffee grounds into polylactic acid using two methods, melt mixing and solvent casting. Alterations to the rheological and thermal behavior of the material were pointed out. However, the mechanical performance of the composites was not part of the study. Here we focus on demonstrating the viability of fully compostable biochar composites using an industrially relevant converting process, such as thermoforming. The processing conditions, as well as the mechanical performance are discussed, paving the way towards the large scale production of consumer products and packaging.

**Table 1.** Selected studies since 2015 reporting biochar–plastic composites.

| Publication Year | Biochar Feedstock | Base Polymer | Citation |
|---|---|---|---|
| 2015/2016 | waste wood (*Pinus radiata*), landfill pine sawdust, sewage sludge, and poultry litter | PP | [9–11] |
| 2016 | bamboo | PE | [12] |
| 2017 | waste coffee | PBAT | [13] |
| 2018 | bamboo | PLA | [14] |
| 2019 | NS | PP | [15] |
| 2019 | waste coffee | PE | [16] |
| 2019 | wheat straw, *Miscanthus*, oilseed rape, rice husk, and mixed softwoods | epoxy | [17] |
| 2019 | sugarcane bagasse | PE | [18] |
| 2019 | rice husk | starch | [19] |
| 2019 | maple wood, waste coffee | epoxy | [20,21] |
| 2019/2020 | rice husk, poplar wood | PE | [22,23] |
| 2020 | waste coffee | PLA | [24] |
| 2020 | *Miscanthus* | PHBV | [25] |
| 2020 | soyhull meal | PP | [26] |
| 2021 | wood, sewage sludge | PLA | [27] |

NS: not specified; PHBV: poly (3-hydroxybutyrate-co-3-hydroxyvalerate); PBAT: poly (butylene adipate-co-terephthalate); PLA: poly (lactic acid); PE: polyethylene; PP: polypropylene.

## 2. Materials and Methods

### 2.1. Materials

Corn starch was obtained from MP Biomedicals LLC (Solon, OH, USA). Polycaprolactone (PCL) Capa 6800 was supplied by Perstorp (Warrington, UK). Biochar was derived from spent coffee grounds obtained from the Rochester Institute of Technology (RIT) cafeteria. The material was first dried using an in-house batch dehydrator (Ecovim-250, Ecovim USA, Los Angeles, CA, USA) and then processed in a commercial-scale "Biogenic Refinery" manufactured by Biomass Controls (Putnam, CT, USA) and owned by RIT [33]. To produce biochar, dried coffee grounds were fed through a hopper and auger assembly at an average flow rate of approximately 5 kg/h. The temperature setpoint of 800 °C was maintained within ±25 °C over the course of the approximately 3-h experiment. After thermochemical conversion, a dual auger system transported the final biochar product to the collection box, where samples were quenched with water to cool the material and prevent further reaction with ambient air.

### 2.2. Sample Preparation

Thermoplastic starch was made using an internal shear mixer, CWB Brabender (South Hackensack, NJ, USA) Intelli-torque Plasticorder torque rheometer with a 60cc 3-piece mixing head. TPS starch was blended at 30, 40, 50, and 60 wt.% with PCL in the mixer at 100 °C for 8 min and 50 rpm. The equilibrium torque was recorded as an indirect measurement of the viscosity of the melt, as shown in Table 2. The samples were compression molded with a heated press (Carver 4391, Wabash, IN, USA). Thermoforming was performed on a Sencorp (Barnstable, MA, USA) Cera TEK 810/1-CE sheetfed laboratory thermoformer using a male mold. Optimum forming conditions were achieved through trial and error by adjusting the heating temperature and dwell time and monitoring the wrapping and webbing in the blisters (see Table 2).

Biochar composites were manufactured using a 50:50 PCL:Starch blend as the base material, with 10, 20, and 30 wt.% biochar mixed at 85 °C. This base material was selected based on the thermoforming ability while maintaining a high elongation at break and starch content. Thermoforming was performed at 138 °C, a temperature significantly higher than that of the material without biochar (50:50 row in Table 2). However, going from 10 to 30 wt.% biochar did not affect the thermoforming temperature.

**Table 2.** Processing conditions for sample preparation.

| Material Composition PCL:Starch | Mixing Equilibrium Torque (Nm) | Temp (°C) | Compression Molding Pressure (tons) | Time (min) | Thermoforming Forming Temp (°C) | Time (min) |
|---|---|---|---|---|---|---|
| 60:40 | 12 | 200 | 3 | 7 | 110 | 1.5 |
| 50:50 | 13 | 200 | 3.5 | 8 | 113 | 1 |
| 40:60 | 17 | 180 | 6.5 | 15 | 116 | 1 |
| 30:70 | 21 | 210 | 7 | 15 | 138 | 1 |

## 2.3. Mechanical Properties Characterization

Tensile testing of the blend was carried out using an Instron (Norwood, MA, USA) Universal Testing Machine model 5567 at a crosshead speed of 12.5 mm/min. At least five specimens of each sample were tested according to American Society for Testing and Materials (ASTM) standard D638. Samples were conditioned at room temperature for at least 24 h before mechanical testing. Type 5 samples were cut from the compression molded sheet with a thickness of approximately 1 mm (similar to the sheet shown in Figure 1).

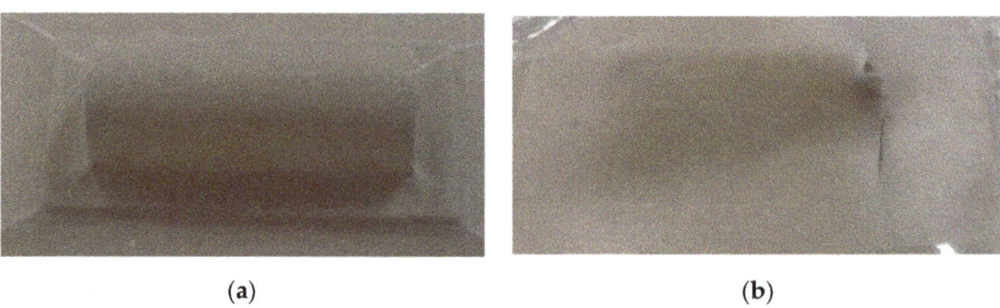

(a)      (b)

**Figure 1.** Thermoformed samples containing 60 wt.% starch (**a**) and 70 wt.% starch (**b**).

## 2.4. Soil Burial Test/Aerobic Biodegradation

Cellulose, PCL60/Starch40, and PCL40/Starch60 samples were cut into 2.54 cm square pieces to obtain a uniform sample size for degradation. Eighteen samples of each specimen were prepared and weighed to record their initial weight. The samples were buried in the soil at a depth of about 2.5 cm. The test was carried out at room temperature (i.e., 22 °C). Water was sprinkled on the soil surface every three days to ensure that the soil remained humid. The samples were measured for weight loss every 7 days from the day they were initially buried. Three samples of each specimen were measured by washing them gently with distilled water and drying the samples at 60 °C in a vacuum oven until a constant weight was obtained. Weight loss percentage was calculated based on Equation (1),

$$\text{Weight loss } (\%) = \frac{w_i - w_d}{w_i} \times 100 \tag{1}$$

where $w_d$ is the dry weight of the film after being washed with distilled water and $w_i$ is the initial dry weight of the specimen [34].

## 3. Results and Discussion

Table 2 shows the processing conditions for the three stages of sample preparation: mixing, compression molding, and thermoforming. As the starch content in the blend increased, the equilibrium torque increased. This indicates that the viscosity of the blend increased due to an increase in the starch content. A higher torque requirement for blending with higher starch content also indicates that a higher pressure was required for the conversion process. This can be evidenced in the increase in pressure requirement for the compression molding stage, accompanied by an increase in temperature. Similarly, an increase in the starch content increased the forming temperature in the thermoformer (see Table 2).

Thermoformed blisters were successfully manufactured with starch contents up to 60 wt.%. Above 60 wt.% starch, the material was unsuitable for thermoforming due to decreased pliability and the blend being too fragile (see Figure 1).

Figure 2 shows the effect of PCL:starch proportions on the mechanical properties. All samples showed typical elastomeric behavior with some degree of strain hardening. Pure PCL had the highest average tensile strength of 55 MPa. The plot displays a U-shape where the strength decreased and then increased at higher starch concentrations (i.e., 70 wt.%). This behavior could indicate an incompatibility of the PCL and starch since the strength of some blends was lower than that of pure PLC and sample with 70 wt.% starch [35]. Similarly, PCL had the highest percentage of elongation at break, which was expected due to its rubbery nature [36]. As the starch content increased, the elongation decreased. However, at 60 wt.%, the elongation was higher than at 50 wt.%. This difference may be attributed to the differences in processing conditions, as shown in Table 2, where the compression molding of the 40:60 sample was done at a lower temperature but higher pressure. This result also points out the sensitivity of the material to processing conditions. Increasing the starch content from 60 to 70 wt.% caused a sharp drop in the elongation from 740% to 26%.

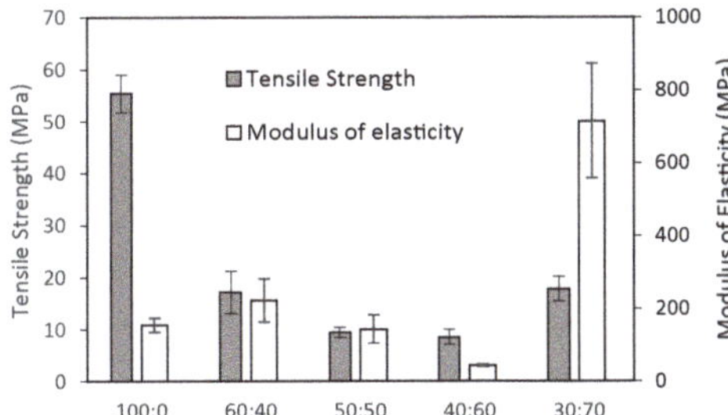
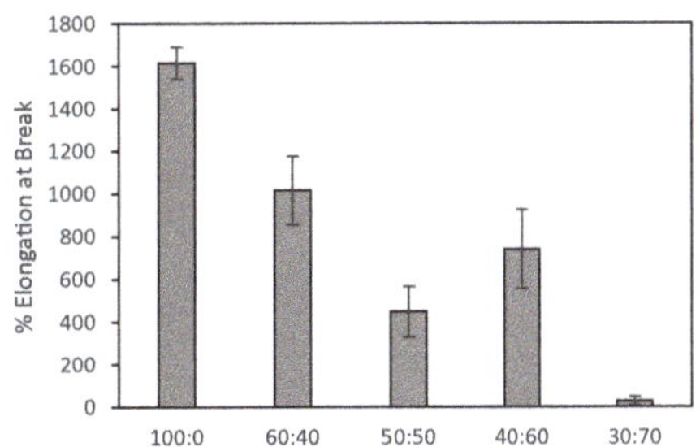

**Figure 2.** Effect of blend ratio of PCL:Starch on (**a**) tensile strength and modulus of elasticity and (**b**) percentage of elongation at break.

The modulus of elasticity was highest at 70 wt.% starch. The stiffness dropped significantly from 70 to 60 wt.% starch. This drop correlates with the difference in conditions for the compression molding stage (see Table 2), which affected both the modulus of elasticity and the elongation at break. A further decrease in the amount of starch showed a nearly linear increase in modulus of elasticity from 60 to 40 wt.% from 43 to 224 MPa, just above the modulus of elasticity of neat PCL (156 MPa).

All the mechanical properties drastically changed, going from 60 to 70 wt.% starch, suggesting a major change in the structure of the blend where PCL is not the majority component, and the properties of starch dictate the properties of the blend. This lack of elongation and high stiffness supports the inability to thermoform the 70 wt.% starch blend.

Figure 3 shows the effect of adding biochar to the TPS containing 50:50 PCL:starch. Adding biochar increased the modulus of elasticity and slightly reduced the tensile strength. Similar results have been observed when reinforcing bioplastics with natural fibers [37]. Varying the biochar content from 10 to 30 wt.% did not have a significant effect on the tensile strength and modulus of elasticity of the material (Figure 3a,b). Conversely, the elongation at break was drastically reduced with the inclusion of biochar. Increasing the amount of biochar from 10 to 30 wt.% further reduced the elongation at break, making the composites significantly more brittle.

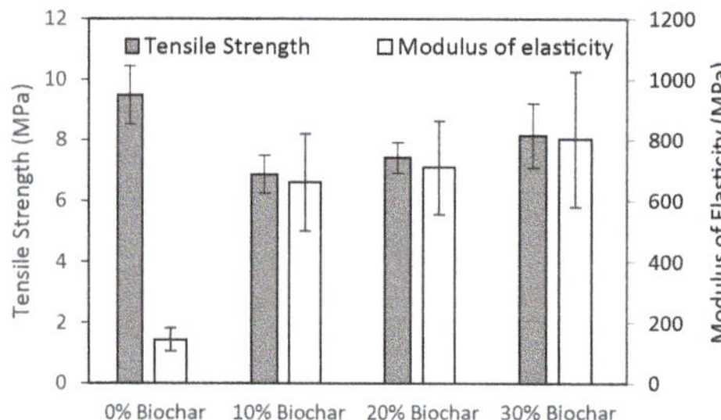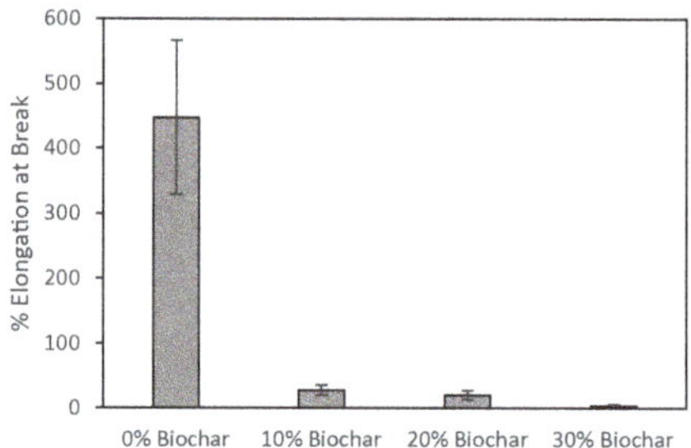

**Figure 3.** Effect of biochar content in 50:50 PCL:starch blend on (**a**) tensile strength and modulus of elasticity and (**b**) percentage of elongation at break.

To demonstrate the thermoforming ability of the composite with biochar, a male mold of a coffee cup lid was manufactured to demonstrate a potential application for this biodegradable composite material. All the composites with biochar allowed the sheet to be thermoformed into coffee cup lids with loadings up to 30 wt.%. Biochar has shown good dispersion in polymeric matrices, such as polypropylene [15] and polylactic acid [24]. It is expected that the biochar composites presented here have a good dispersion given the high shear melt mixing process used. Figure 4 shows a coffee lid containing 10 wt.% biochar. Increasing the biochar load did not affect the thermoforming ability; however, the surface was rougher with less resolution of the details of the mold. The results demonstrate the potential to use biochar as a filler material in thermoform containers and packaging. Additionally, this is an example of a product for coffee shops made from their own waste (i.e., spent coffee grounds). Biochar thus may offer an opportunity for a close-loop economy while displacing plastic or creating fully biodegradable solutions.

**Figure 4.** Thermoformed coffee lid made with 10 wt.% biochar from spent coffee grounds.

Ongoing research is looking at structure–property relationships to better understand the changes observed here. Additionally, the rheology of the material should be further studied to expand the findings of this research to other conversion processes, such as injection molding and blown film extrusion.

Finally, Figure 5 shows the biodegradation of two samples containing 40 and 60 wt.% starch. Higher starch content resulted in a higher level of degradation. These results agree with previous studies [2]. No literature was found on the effect of biochar on the biodegradation of biochar composites. Our previous study showed similar or better biodegradation under anaerobic conditions when calcium carbonate was used as a filler in polylactic acid [38,39]. Preliminary experiments suggest that the addition of biochar enhances biodegradation and further experimentation is ongoing.

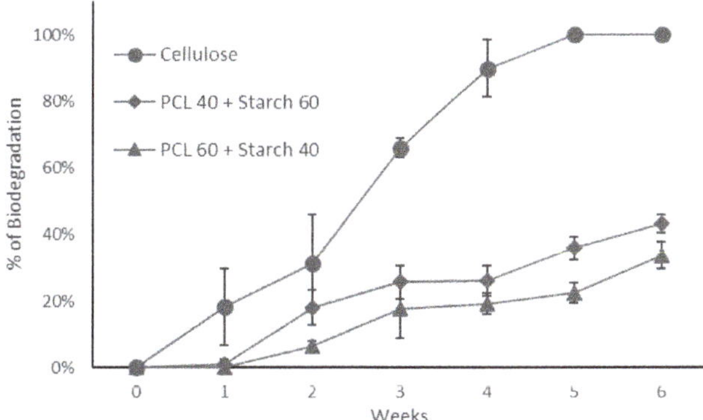


**Figure 5.** Cumulative biodegradation of selected samples and a positive (cellulose) control.

## 4. Conclusions

This study demonstrated the thermoforming of fully biodegradable thermoplastic starch without the use of water or glycerol but instead a rubbery biopolymer (i.e., PCL). The ratio of PCL and starch affected the processing conditions as well as the mechanical properties. Thermoformed blisters were successfully made with starch contents from 30 to 60 wt.%. Increasing the starch content beyond that point drastically changed the properties and rendered the material unsuited for thermoforming. Biochar composites were made using the 50:50 PCL:starch material. Prototype thermoformed coffee lids were made with content up to 30 wt.% biochar derived from waste coffee grounds. Manufacturing of composites using biochar demonstrates the possibility to manufacture fully biodegradable items and the valorization of a byproduct from the pyrolysis of organic waste, thus promoting a circular economy model for future sustainable packaging products.

**Author Contributions:** Conceptualization, K.D., T.A.T. and C.A.D.; methodology, T.A.T. and C.A.D.; software, C.A.D.; validation, R.K.S., T.E. and C.A.D.; formal analysis, R.K.S., T.E., T.A.T., K.D. and C.A.D.; investigation, R.K.S., T.E., T.A.T., K.D. and C.A.D.; resources, K.D., T.A.T. and C.A.D.; data curation, R.S., T.E. and C.A.D.; writing—original draft preparation, R.K.S., T.E., T.A.T., K.D. and C.A.D.; writing—review and editing, K.D., T.A.T. and C.A.D.; visualization, R.K.S., T.E., T.A.T. and C.A.D.; supervision, C.A.D.; project administration, C.A.D.; funding acquisition, C.A.D. All authors have read and agreed to the published version of the manuscript.

**Funding:** This research was funded by the Center for Sustainable Packaging at the Rochester Institute of Technology.

**Conflicts of Interest:** The authors declare no conflict of interest.

## References

1. Lu, D.R.; Xiao, C.M.; Xu, S.J. Starch-based completely biodegradable polymer materials. *Express Polym. Lett.* **2009**, *3*, 366–375. [CrossRef]
2. Nunziato, R.; Hedge, S.; Dell, E.; Trabold, T.; Lewis, C.; Diaz, C. Mechanical Properties and Anaerobic Biodegradation of Thermoplastic Starch/Polycaprolactone Blends. In Proceedings of the 21th IAPRI World Conference on Packaging, Zhuhai, China, 19–22 June 2018; pp. 722–729.
3. Averous, L.; Moro, L.; Dole, P.; Fringant, C. Properties of thermoplastic blends: Starch–polycaprolactone. *Polymer* **2000**, *41*, 4157–4167. [CrossRef]
4. Lopez, O.; Garcia, M.A.; Villar, M.A.; Gentili, A.; Rodriguez, M.S.; Albertengo, L. Thermo-compression of biodegradable thermoplastic corn starch films containing chitin and chitosan. *LWT Food Sci. Technol.* **2014**, *57*, 106–115. [CrossRef]
5. Zabaniotou, A.; Rovas, D.; Libutti, A.; Monteleone, M. Boosting circular economy and closing the loop in agriculture: Case study of a small-scale pyrolysis-biochar based system integrated in an olive farm in symbiosis with an olive mill. *Environ. Dev.* **2015**, *14*, 22–36. [CrossRef]
6. Chen, W.; Meng, J.; Han, X.; Lan, Y.; Zhang, W. Past, present, and future of biochar. *Biochar* **2019**, *1*, 75–87. [CrossRef]

7. Weber, K.; Quicker, P. Properties of biochar. *Fuel* **2018**, *217*, 240–261. [CrossRef]
8. Guran, S. Sustainable waste-to-energy technologies: Gasification and pyrolysis. In *Sustainable Food Waste-to-Energy Systems*; Elsevier: Amsterdam, The Netherlands, 2018; pp. 141–158, ISBN 9780128111574.
9. Das, O.; Sarmah, A.K.; Bhattacharyya, D. A novel approach in organic waste utilization through biochar addition in wood/polypropylene composites. *Waste Manag.* **2015**, *38*, 132–140. [CrossRef]
10. Das, O.; Sarmah, A.K.; Bhattacharyya, D. Biocomposites from waste derived biochars: Mechanical, thermal, chemical, and morphological properties. *Waste Manag.* **2016**, *49*, 560–570. [CrossRef]
11. Das, O.; Bhattacharyya, D.; Sarmah, A.K. Sustainable eco-composites obtained from waste derived biochar: A consideration in performance properties, production costs, and environmental impact. *J. Clean. Prod.* **2016**, *129*, 159–168. [CrossRef]
12. Li, S.; Li, X.; Chen, C.; Wang, H.; Deng, Q.; Gong, M.; Li, D. Development of electrically conductive nano bamboo charcoal/ultra-high molecular weight polyethylene composites with a segregated network. *Compos. Sci. Technol.* **2016**, *132*, 31–37. [CrossRef]
13. Moustafa, H.; Guizani, C.; Dufresne, A. Sustainable biodegradable coffee grounds filler and its effect on the hydrophobicity, mechanical and thermal properties of biodegradable PBAT composites. *J. Appl. Polym. Sci.* **2017**, *134*. [CrossRef]
14. Qian, S.; Yan, W.; Zhu, S.; Fontanillo Lopez, C.A.; Sheng, K. Surface modification of bamboo-char and its reinforcement in PLA biocomposites. *Polym. Compos.* **2018**, *39*, E633–E639. [CrossRef]
15. Abdelwahab, M.A.; Rodriguez-Uribe, A.; Misra, M.; Mohanty, A.K. Injection Molded Novel Biocomposites from Polypropylene and Sustainable Biocarbon. *Molecules* **2019**, *24*, 4026. [CrossRef] [PubMed]
16. Arrigo, R.; Jagdale, P.; Bartoli, M.; Tagliaferro, A.; Malucelli, G. Structure-property relationships in polyethylene-based composites filled with biochar derived from waste coffee grounds. *Polymers* **2019**, *11*, 1336. [CrossRef]
17. Bartoli, M.; Giorcelli, M.; Rosso, C.; Rovere, M.; Jagdale, P.; Tagliaferro, A. Influence of Commercial Biochar Fillers on Brittleness/Ductility of Epoxy Resin Composites. *Appl. Sci.* **2019**, *9*, 3109. [CrossRef]
18. Ferreira, G.F.; Pierozzi, M.; Fingolo, A.C.; da Silva, W.P.; Strauss, M. Tuning Sugarcane Bagasse Biochar into a Potential Carbon Black Substitute for Polyethylene Composites. *J. Polym. Environ.* **2019**, *27*, 1735–1745. [CrossRef]
19. George, J.; Azad, L.B.; Poulose, A.M.; An, Y.; Sarmah, A.K. Nano-mechanical behaviour of biochar-starch polymer composite: Investigation through advanced dynamic atomic force microscopy. *Compos. Part A Appl. Sci. Manuf.* **2019**, *124*, 105486. [CrossRef]
20. Giorcelli, M.; Khan, A.; Pugno, N.M.; Rosso, C.; Tagliaferro, A. Biochar as a cheap and environmental friendly filler able to improve polymer mechanical properties. *Biomass Bioenergy* **2019**, *120*, 219–223. [CrossRef]
21. Giorcelli, M.; Bartoli, M. Development of Coffee Biochar Filler for the Production of Electrical Conductive Reinforced Plastic. *Polymers* **2019**, *11*, 1916. [CrossRef]
22. Zhang, Q.; Khan, M.U.; Lin, X.; Cai, H.; Lei, H. Temperature varied biochar as a reinforcing filler for high-density polyethylene composites. *Compos. Part B Eng.* **2019**, *175*, 107151. [CrossRef]
23. Zhang, Q.; Zhang, D.; Xu, H.; Lu, W.; Ren, X.; Cai, H.; Lei, H.; Huo, E.; Zhao, Y.; Qian, M.; et al. Biochar filled high-density polyethylene composites with excellent properties: Towards maximizing the utilization of agricultural wastes. *Ind. Crops Prod.* **2020**, *146*, 112185. [CrossRef]
24. Arrigo, R.; Bartoli, M.; Malucelli, G. Poly(lactic Acid)-biochar biocomposites: Effect of processing and filler content on rheological, thermal, and mechanical properties. *Polymers* **2020**, *12*, 892. [CrossRef] [PubMed]
25. Li, Z.; Reimer, C.; Wang, T.; Mohanty, A.K.; Misra, M. Thermal and Mechanical Properties of the Biocomposites of Miscanthus Biocarbon and Poly(3-Hydroxybutyrate-co-3-Hydroxyvalerate) (PHBV). *Polymers* **2020**, *12*, 1300. [CrossRef] [PubMed]
26. Watt, E.; Abdelwahab, M.A.; Snowdon, M.R.; Mohanty, A.K.; Khalil, H.; Misra, M. Hybrid biocomposites from polypropylene, sustainable biocarbon and graphene nanoplatelets. *Sci. Rep.* **2020**, *10*, 10714. [CrossRef] [PubMed]
27. Pudełko, A.; Postawa, P.; Stachowiak, T.; Malińska, K.; Dróżdż, D. Waste derived biochar as an alternative filler in biocomposites—Mechanical, thermal and morphological properties of biochar added biocomposites. *J. Clean. Prod.* **2021**, *278*, 123850. [CrossRef]

28. Sarasini, F.; Luzi, F.; Dominici, F.; Maffei, G.; Iannone, A.; Zuorro, A.; Lavecchia, R.; Torre, L.; Carbonell-Verdu, A.; Balart, R.; et al. Effect of different compatibilizers on sustainable composites based on a PHBV/PBAT matrix filled with coffee silverskin. *Polymers* **2018**, *10*, 1256. [CrossRef]
29. García-García, D.; Carbonell, A.; Samper, M.D.; García-Sanoguera, D.; Balart, R. Green composites based on polypropylene matrix and hydrophobized spend coffee ground (SCG) powder. *Compos. Part B Eng.* **2015**, *78*, 256–265. [CrossRef]
30. Wu, C.S. Renewable resource-based green composites of surface-treated spent coffee grounds and polylactide: Characterisation and biodegradability. *Polym. Degrad. Stab.* **2015**, *121*, 51–59. [CrossRef]
31. Zarrinbakhsh, N.; Wang, T.; Rodriguez-Uribe, A.; Misra, M.; Mohanty, A.K. Characterization of wastes and coproducts from the coffee industry for composite material production. *BioResources* **2016**, *11*, 7637–7653. [CrossRef]
32. Quosai, P.; Anstey, A.; Mohanty, A.K.; Misra, M. Characterization of biocarbon generated by high- and low-temperature pyrolysis of soy hulls and coffee chaff: For polymer composite applications. *R. Soc. Open Sci.* **2018**, *5*. [CrossRef]
33. Rodriguez Alberto, D.; Repa, K.; Hegde, S.; Miller, C.W.; Trabold, T.A. Novel Production of Magnetite Particles via Thermochemical Processing of Digestate from Manure and Food Waste. *IEEE Magn. Lett.* **2019**. [CrossRef]
34. Phetwarotai, W.; Potiyaraj, P.; Aht-Ong, D. Biodegradation of Polylactide and Gelatinized Starch Blend Films Under Controlled Soil Burial Conditions. *J. Polym. Environ.* **2013**, *21*, 95–107. [CrossRef]
35. Weng, F.; Zhang, P.; Guo, D.; Koranteng, E.; Wu, Z.; Wu, Q. Preparation and Properties of Compatible Starch-PCL Composites: Effects of the NCO Functionality in Compatibilizer. *Starch Stärke* **2020**, *72*, 1900239. [CrossRef]
36. Liang, J.Z.; Duan, D.R.; Tang, C.Y.; Tsui, C.P.; Chen, D.Z. Tensile properties of PLLA/PCL composites filled with nanometer calcium carbonate. *Polym. Test.* **2013**, *32*, 617–621. [CrossRef]
37. Chaiwong, W.; Samoh, N.; Eksomtramage, T.; Kaewtatip, K. Surface-treated oil palm empty fruit bunch fiber improved tensile strength and water resistance of wheat gluten-based bioplastic. *Compos. Part B Eng.* **2019**, *176*, 107331. [CrossRef]
38. Hegde, S.; Dell, E.; Lewis, C.; Trabold, T.A.; Diaz, C.A. Anaerobic Biodegradation of Bioplastic Packaging Materials. In Proceedings of the 21st IAPRI World Conference on Packaging, Zhuhai, China, 19–22 June 2018; DEStech Publications, Inc.: Lancaster, PA, USA, 2018; pp. 730–737.
39. Hegde, S.; Dell, E.M.; Lewis, C.L.; Trabold, T.A.; Diaz, C.A. Development of a biodegradable thermoformed tray for food waste handling. In Proceedings of the 29th IAPRI, Enschede, The Netherlands, 11–14 June 2019; pp. 553–561.

**Publisher's Note:** MDPI stays neutral with regard to jurisdictional claims in published maps and institutional affiliations.

*Review*

# Hydrochars as Emerging Biofuels: Recent Advances and Application of Artificial Neural Networks for the Prediction of Heating Values

Ioannis O. Vardiambasis [1], Theodoros N. Kapetanakis [1], Christos D. Nikolopoulos [1], Trinh Kieu Trang [2], Toshiki Tsubota [3], Ramazan Keyikoglu [4], Alireza Khataee [4,5] and Dimitrios Kalderis [1,*]

[1]   Department of Electronic Engineering, Hellenic Mediterranean University, Chania, 73100 Crete, Greece; ivardia@hmu.gr (I.O.V.); todokape@hmu.gr (T.N.K.); cnikolo@hmu.gr (C.D.N.)
[2]   Applied Chemistry Course, Department of Engineering, Graduate School of Engineering, Kyushu Institute of Technology, 1-1 Sensuicho, Tobata-ku, Kitakyushu 804-8550, Japan; trinhkieutrang_t57@hus.edu.vn
[3]   Department of Applied Chemistry, Faculty of Engineering, Kyushu Institute of Technology, 1-1 Sensuicho, Tobata-ku, Kitakyushu 804-8550, Japan; tsubota@che.kyutech.ac.jp
[4]   Department of Environmental Engineering, Gebze Technical University, 41400 Gebze, Turkey; ramazankeyikoglu@gmail.com (R.K.); a_khataee@tabrizu.ac.ir (A.K.)
[5]   Research Laboratory of Advanced Water and Wastewater Treatment Processes, Department of Applied Chemistry, Faculty of Chemistry, University of Tabriz, Tabriz 51666-16471, Iran
*   Correspondence: kalderis@hmu.gr; Tel.: +30-28210-23017

Received: 19 August 2020; Accepted: 1 September 2020; Published: 3 September 2020

**Abstract:** In this study, the growing scientific field of alternative biofuels was examined, with respect to hydrochars produced from renewable biomasses. Hydrochars are the solid products of hydrothermal carbonization (HTC) and their properties depend on the initial biomass and the temperature and duration of treatment. The basic (Scopus) and advanced (Citespace) analysis of literature showed that this is a dynamic research area, with several sub-fields of intense activity. The focus of researchers on sewage sludge and food waste as hydrochar precursors was highlighted and reviewed. It was established that hydrochars have improved behavior as fuels compared to these feedstocks. Food waste can be particularly useful in co-hydrothermal carbonization with ash-rich materials. In the case of sewage sludge, simultaneous P recovery from the HTC wastewater may add more value to the process. For both feedstocks, results from large-scale HTC are practically non-existent. Following the review, related data from the years 2014–2020 were retrieved and fitted into four different artificial neural networks (ANNs). Based on the elemental content, HTC temperature and time (as inputs), the higher heating values (HHVs) and yields (as outputs) could be successfully predicted, regardless of original biomass used for hydrochar production. ANN$_3$ (based on C, O, H content, and HTC temperature) showed the optimum HHV predicting performance ($R^2$ 0.917, root mean square error 1.124), however, hydrochars' HHVs could also be satisfactorily predicted by the C content alone (ANN$_1$, $R^2$ 0.897, root mean square error 1.289).

**Keywords:** hydrochar; hydrothermal carbonization; CiteSpace; scientometric analysis; artificial neural network; biofuels

---

## 1. Introduction

Subcritical water is hot water (100–374 °C) under enough pressure to maintain its liquid state. At these conditions, the dielectric constant of water is reduced, therefore it becomes a good solvent for non-polar substances. Throughout the 1980–1990s, this property was thoroughly exploited in subcritical water extraction and chromatography, for the isolation of natural products from environmental matrices and the replacement of hazardous organic solvents, respectively [1–5]. The tunable properties of subcritical water were later utilized to degrade organic contaminants in wastewater and soils. Several groups showed that recalcitrant contaminants, such as explosives and pesticides, can be degraded in-situ in relatively short times [6,7]. Furthermore, the addition of the environmentally-friendly hydrogen peroxide in subcritical water accelerates the degradation of contaminants due to the production of the highly reactive hydroxyl radicals [8,9].

The term 'hydrothermal carbonization' (HTC) started appearing in the literature regularly in the early 2000s, to describe the upgrading or modification of materials and the synthesis of nanostructures in a subcritical water environment [10–12]. It still remains one of the main methods for the production of nanosized inorganic materials. Hydrothermal carbonization of biomass for the production of hydrochars for fuel purposes was first reported in 2010 [13]. In an effort to develop alternative fuels from sustainable sources, researchers focused on residual biomasses and agricultural by-products as hydrochar feedstocks. Since then, the number of papers that have studied hydrochar as biofuel has been steadily increasing. Depending on location and availability, a large number of biomasses has been investigated, from food waste and bamboo dust, to poultry litter and sugarcane bagasse. The main advantage—and difference from dry pyrolysis—is the method's potential to process high moisture biomasses. In all cases, the objectives were common: a competitive higher heating value (HHV) and a high solid yield. The mechanisms of biomass conversion to hydrochar have been established and reviewed in the literature [14–16]. Dehydration, decarboxylation, and decarbonylation reactions occur, the extent of which depends on processing conditions (mainly temperature, treatment time, and pH of feed water).

A few empirical models that provide the hydrochar mass yield (%) and HHVs have been proposed [17,18]. However, these models are highly dependent on the biomass used and are therefore of limited applicability. Some others require an increased number of laboratory analyses for model input. The model developed by Conag et al. (2018) focused on sugarcane bagasse only and its charred derivatives. The suggested equation provided an adequate estimate of the HHV having a mean absolute error of 6.1% and a coefficient of determination (r2) of 0.91 [19]. Based on the severity factor ($R_o$), polarity index (IP), and reactivity index (IR), Vallejo et al. (2020) developed a multilinear model for the prediction of the HHVs of hydrochars from various biomasses. However, the difficulty in using IP and IR is that earlier determination of C, H, N, S, hemicellulose, aqueous extractives, lignin, and ash content in the raw biomass is required [20]. Similarly, the regression model proposed by Akdeniz et al. (2020) required a significant number of time-consuming laboratory analyses as input data [21]. Furthermore, due to the complexity of lignocellulosic biomass even a small change in the experimental conditions (e.g., such as moisture content) may result in considerably different yields and/or HHVs. Additionally, different types of hydrothermal carbonization reactors have different heat transfer values, which affect the reaction rates and subsequently the composition of the final solid product. To date, an accurate and generic model to correlate HHVs to the very basic hydrochar properties, regardless of initial feedstock, moisture content, and reactor size/type, is missing [18].

Artificial intelligence (AI) is a wide area of rapid growth with a large number of applications including but not limited to telecommunications, medical diagnosis, healthcare, and robotics [22,23]. A powerful section of AI is the Artificial Neural Networks (ANNs), the function of which is inspired by the biological central nervous system. In general, ANNs can be treated as a computational method attempting to simulate the complex functions of the human brain. ANNs, and in particular multilayer perception ANNs (MLP-ANNs) with at least 2 hidden layers, can theoretically approximate any nonlinear function between their input and output data and may be considered as universal approximators. The fundamental building block of any ANN is the artificial neuron, which is a simplified form of the biological neuron. Every ANN can be modeled as a layered structure of neurons. The structure is composed of different layers, such as the input layer, a number of intermediate layers which are called hidden layers, and the output layer. The number of hidden layers varies and depends on the problem at hand. Each layer consists of a number of neurons, which through variable synaptic weights are connected to the neurons of the next layer. In addition to synaptic weights, the neurons consist of activation functions which limit the amplitude range of the neurons' outputs to [0,1] or [−1,1]. Something that is also important, in any neuron, is the role of the so-called bias, which is a constant value determining whether the neuron is activated or not. Synaptic weights and biases are free parameters that are calculated through a learning algorithm, in order to achieve the desired target outcome.

The use of ANNs in biomass exploitation studies is still at an early stage but the interest is growing. Bhange et al. (2017) developed a feed forward backpropagation ANN for the garden biomass pretreatment process from experimental data. The results of the developed ANN model were compared to those of the response surface methodology (RSM), achieving a mean square error (MSE) value equal to 0.121 [24]. Baruah et al. (2017) also modeled the same ANN architecture but for biomass gasification in fixed bed downdraft gasifiers. The corresponding ANN outputs for the concentration $CH_4\%$, $CO\%$, $CO_2\%$, and $H_2\%$ gas species were found to be in good agreement with the experimental data, attaining absolute fraction of variance ($R^2$) values higher than 0.98 and root mean square error (RMSE) values less than 0.0915 [25]. Nasrudin et al. (2019) compared various training algorithms for modelling microwave pyrolysis of oil palm fiber for hydrogen and biochar production. Their inputs were the temperature, the microwave power, the nitrogen flow rate and their outputs were the weights of hydrogen and biochar. The best performance was achieved, as expected, by the Levenberg–Marquardt (LM) and the Bayesian Regulation (BR) training algorithms. The LM (BR) algorithm achieved RMSE values equal to 0.206 (0.216) for hydrogen weight and 0.822 (0.886) for biochar weight prediction [26].

Several laboratory studies have investigated the effect of each HTC parameter on hydrochar properties and the behavior of the solid fuel during combustion/incineration. However, HTC is a complex process, largely feedstock-dependent, therefore, homogeneity and standardization of results are still lacking [15,27]. Very few works have developed mathematical models to correlate specific properties of hydrochars to their elemental content and HTC conditions [28]. Based on the above, the objectives of this work were the following: (1) to address the current areas of intense activity and trends of 'hydrochars as fuel' research by a visual scientometrics analysis performed by CiteSpace software, (2) review the recent advances in the areas (clusters) of highest activity, as indicated by the CiteSpace analysis and (3) perform an ANN statistical analysis to correlate the minimum number of fundamental hydrochar properties (regardless of original feedstock and moisture content) to the heating values reported in the published literature of 2014–2020.

The Java-based software CiteSpace was developed by Chaomei Chen in 2006 and it focuses on finding critical points in the development of a field or a domain, including identifying fast-growing topical areas, finding citation hotspots in the land of publications, decomposing a network (of publications, or authors, or geographical areas etc.) into clusters and automatically labeling clusters with the most frequent terms from citing articles [29,30]. The effectiveness of this approach has been shown in different fields, for example on climate change and tourism and recently in emerging trends of biochar research and applications [31,32].

## 2. Data Acquisition, Methods, and Review of Recent Literature

### 2.1. Data Acquisition and Methods

Scopus (2014–2020) was selected as the scientific database and the keywords were 'hydrochar' and 'fuel' (article title, abstract, keywords). A total of 270 papers were retrieved and categorized as follows: articles (225), conference papers (27), reviews (9), book chapters (5), and conference review (1). To gain an insight into the latest and most active research sub-topics, data acquisition was limited to the years 2018–2020. This yielded a total of 175 papers (8506 cited references), which were saved in ris format, as required by CiteSpace (version: 5.6.R5). The software can figure out the relationship between authors and the correlation between keywords as well as point out the emerging trends, hot topics (clusters) and gaps in the 'hydrochar as fuel' research field. In the generated network maps, each node represents one item (e.g., keyword or author), and the size of the node indicates the frequency of this item. The log-likelihood algorithm (LLR) was used as the calculation method to obtain the clustering results [30,31].

With respect to the ANN analysis, Scopus publications from the years 2014–2020 were retrieved, using the same keywords as above. The required input data were the following: temperature (°C) and time (hr) during hydrothermal carbonization, carbon, oxygen, and hydrogen content of hydrochars. The output parameters were the higher heating value (HHV, MJ/kg) and the % solid yield (mass of produced hydrochar/mass of original biomass × 100). These were the most commonly reported parameters in the related published literature. The reactor pressure during treatment was not included because it is known that during HTC, the effect of pressure on the products' composition and yield is minimal [6,8,14]. Other input (e.g., moisture of biomass) and output parameters (e.g. lower heating value) were considered but excluded due to the limited number of studies that have reported such data. From a total of 270 documents, 144 reported full sets of the required data therefore these studies were used for the ANN analysis. All simulations were performed in MATLAB environment, using the deep learning toolbox.

### 2.2. Basic Characteristics of the Reviewed Publications (2014–2020)

Figure 1 shows the number of publications and total citations each year for the period of 2014–2020. A steady increase can be observed for both, indicating an active and dynamic field of research. This research field is highly interdisciplinary, since biomass processing, hydrothermal carbonization and hydrochar characterization and application are based on distinctly different knowledge backgrounds. This is represented in the document distribution by subject area, with Energy, Environmental Science, and Chemical Engineering each hosting 29.6, 23.2, and 17.2% of the documents, respectively.

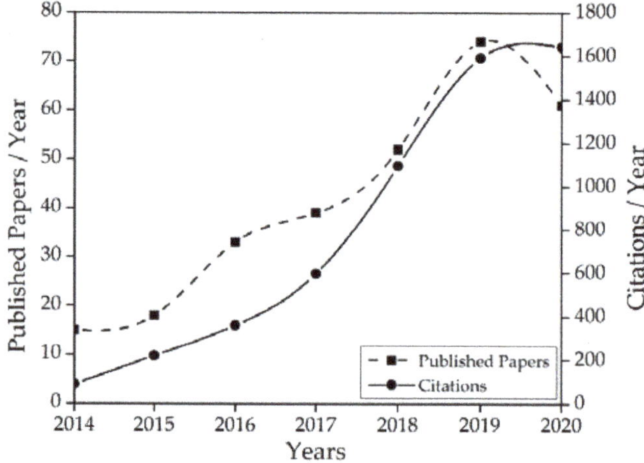


**Figure 1.** Citations and number of papers published using the keywords 'hydrochar' and 'fuel' for the period 2014–2020.

### 2.3. CiteSpace Recent Scientometric Analysis (2018–2020)

Table 1 shows the 8 sub-fields (clusters #0–7) of intense activity determined for the period 2018–2020. Cluster labels are selected from noun phrases and index terms of citing articles (nodes) of each cluster. These terms are ranked by three different algorithms, thoroughly explained in the developer's publications [29,30]. The cluster with the lowest #number corresponds to the sub-field with the highest number of published papers. Cluster #0 (solid fuel hydrochar) corresponds to the fuel properties and combustion behavior of hydrochars, therefore is naturally where most papers have focused. Cluster #1 (sludge-derived hydrochar), #4 (food waste), and #7 (corn stalk) point out the feedstock materials mostly used for fuel hydrochar production. These three clusters account for >50% of the total published papers for the period 2018–2020. The compositional variety of each feedstock greatly adds to the complexity of hydrothermal carbonization and consequently leads to hydrochars with different physicochemical properties. In the following sections, the recent advances with respect to the valorization of sewage sludge and food waste for the production of hydrochar will be reviewed. Cluster #2 (water source) highlights a very important aspect with respect to the techno-economic feasibility of hydrochar production at a large scale. Since water is an essential component during the process, the minimum required quantity will determine the need for potential recycling of HTC wastewater, thus offering an option to improve the overall efficiency. Clusters #3 and 6 (and the papers within) focus on fundamental aspects of hydrothermal carbonization, already thoroughly reviewed in the literature. Cluster #5 relates to the production of gases (greenhouse and others) during the combustion of hydrochars and will form part of a separate study in the future.

The average modularity value, the silhouette, and the most frequent common terms among the nodes of the same cluster can also be seen in Table 1. The modularity of a network of publications measures the extent to which the publications can be decomposed to multiple components or modules. If a network's modularity is close to 1.00, then the network is clearly divided into thematically distinct clusters. In contrast, if modularity is below 0.30, many between-cluster links could be expected. The value of 0.4122 indicates that clusters are rather closely related to each other and there is a certain number of common methodologies that are followed. It also highlights a research field with several gaps and unknown parameters, the effect of which has not been fully investigated. The silhouette value shows the homogeneity of a cluster and takes values between −1 and 1. The higher the silhouette value, the more focused and consistent the papers of this cluster are, provided the clusters in comparison have similar sizes. The silhouette value for each and every cluster showed a significant degree of homogeneity, a result that is more meaningful for the clusters with the highest number of papers.

**Table 1.** Network modularity, cluster silhouette, the number of papers and the most frequently reported terms in the 2018–2020 clusters.

| Network Modularity: 0.4122 | | | |
|---|---|---|---|
| Cluster | Silhouette | Number of Papers | Most Frequently Reported Common Terms Among the Papers of Each Cluster [a] |
| 0 (solid fuel hydrochar) | 0.534 | 61 | microwave, synthesis, green waste, fuel properties |
| 1 (sludge-derived hydrochar) | 0.578 | 47 | sustainable biomass fuel, sewage sludge, pelletization technique |
| 2 (water source) | 0.467 | 40 | hydrochar properties, correlations, orange peel waste, chemical constitution |
| 3 (hydrothermal liquid product) | 0.648 | 38 | pyrolysis behaviour, pinewood sawdust, maize straw |
| 4 (food waste) | 0.59 | 37 | comprehensive investigation, water source, food waste, energy potential |
| 5 (gas emission) | 0.639 | 27 | solid biofuel production, effects, biogas generation, co-hydrothermal gasification |
| 6 (physicochemical properties) | 0.841 | 15 | faecal sludge treatment, molasses utilization, alternative solid fuel, gas emissions |
| 7 (corn stalk) | 0.847 | 13 | combustion kinetics, chilean biomass residues, corn stalk |

[a] The terms 'hydrothermal carbonization' and 'hydrochar' appear in all 8 clusters and were not added.

The most commonly encountered keywords are presented in Table 2. Centrality is a relative term that quantifies the importance of a keyword within the research network. The keywords with high frequency and high centrality are generally considered as key nodes, indicating that they have a strong influence in the whole network for the given period [30]. All keywords of Table 2 are closely related to our research field and to each other, whereas expectedly 'temperature' showed the highest centrality value since it is the most influential parameter during hydrothermal carbonization [14,15].

**Table 2.** The top 10 keywords related to 'hydrochar as fuel' research field for the years 2018–2020.

| Rank | Keyword | Frequency | Centrality |
|------|---------|-----------|------------|
| 1 | Hydrothermal carbonization | 134 | 0.01 |
| 2 | Carbonization | 132 | 0.01 |
| 3 | Thermochemistry | 109 | 0.03 |
| 4 | Hydrochar | 85 | 0.05 |
| 5 | Fuel | 74 | 0.07 |
| 6 | Carbon | 63 | 0.03 |
| 7 | Biome | 57 | 0.02 |
| 8 | Combustion | 54 | 0.07 |
| 9 | Temperature | 46 | 0.12 |
| 10 | Calorific value | 37 | 0.01 |

### 2.3.1. Valorization of Sewage Sludge for the Production of Fuel Hydrochar

Sewage sludge management remains in the center of attention throughout the world. In the European Union, new directives and regulations focus on stabilization and valorization options instead of storage routes, such as landfilling or back-filling of mining areas [33–35]. Among the first to investigate the use of sewage sludge for hydrochar production, Zhao et al. (2014) realized that HTC has the advantages of volume reduction and energy densification of the original biomass. The authors concluded that temperature was the most influential parameter and suggested moderate HTC conditions (200 °C, 30 min treatment time) to produce hydrochar with an energy recovery rate of 50% [36]. The results of Kim et al. (2014) largely agreed with these conclusions and further established the improvement in the fuel-related properties of hydrochars compared to the sewage sludge feedstock [37]. At 220 °C, a higher heating value of 18.3 MJ/kg was achieved, comparable to that of lignite coal and 12% higher than that of raw sewage sludge. Several researchers have confirmed the upgrading of fuel quality of sewage sludge through HTC, the categorization of hydrochar in the region of lignite (HHV 15–25 MJ/kg) and the role of temperature as the main influential parameter [28,38–41]. However, most of these studies also showed gradual increases of ash content as the HTC temperature was raised, not a desirable attribute for fuel applications. It is therefore essential that the generally high ash content in sewage sludge (compared to lignocellulosic biomasses) is further investigated as it may affect the ash fusion temperature and slagging potential of hydrochar during combustion. The relative composition of ash determines to a large extent its behavior during combustion: potassium and sodium are transferred to the wastewater during HTC, whereas magnesium, phosphorus, and calcium mostly remain in the resultant solid fuel [42–45]. Therefore, if Mg and Ca minerals dominate the ash fraction in sewage sludge, they will be bound to the hydrochar matrix, thus increasing the possibility of causing slagging and fouling during combustion. The crucial role of ash and the transformation routes of sewage sludge building blocks (lipids, proteins, and polysaccharides) during HTC have been established and thoroughly discussed [46–48].

Recently, hydrothermal co-carbonization (co-HTC) has attracted attention in an effort to reduce the ash content, increase yields and generally improve the fuel properties of the final product. Ma et al. (2019a, 2019b), examined the pyrolysis and gasification behavior of hydrochars prepared by co-HTC of sewage sludge and sawdust. Their thermodynamic and kinetics assessment indicated that co-HTC improved the pyrolysis reactivity and devolatilization performance of sewage sludge hydrochar [49]. Furthermore, the addition of sawdust increased the fixed carbon and calorific value of the produced hydrochars, whereas their gasification runs resulted in a maximized CO content at the optimum sawdust/sewage sludge ratio of 0.25 [50]. However, since both sewage sludge and sawdust were dried before HTC, water was added manually to maintain the hydrothermal conditions. To improve the water efficiency, partially dewatered sludge could be used and the moisture content of the mixture controlled by addition of various quantities of dry sawdust. The same co-HTC rationale was followed by Song et al. (2019) who mixed sewage sludge with lignite coal at 1:1 ratio and processed them hydrothermally at the temperature range of 120–300 °C. The authors determined that the fuel properties of hydrochars were gradually enhanced as the temperature was raised and the optimum HHV of 16.93 MJ/kg was achieved at 240 °C and 30 min residence time [51]. In a similar approach, Wang et al. (2020) mixed dewatered sewage sludge with phenolic wastewater and produced hydrochars with increased HHV. This approach led to a substantial increase of hydrochar yield (1.83–31.11%), a higher heating value (1.01–10.01%) and a considerable decrease of ash content (1.39–25.68%), depending on processing temperature and phenol concentration [52]. An advantage of this method was that no fresh water was used since the wet conditions were controlled by the wastewater addition. Promising results have also been obtained through the co-HTC of sewage sludge with microalgae [53], cow dung [54], and food waste [55,56] in the temperature range of 200–230 °C and 30 min residence time.

Worldwide, there is an increasing demand for plant nutrients such as phosphorus (P). The increasing pressure on fossil P sources has directed efforts to recover P from renewable sources. Consequently, some groups have focused on obtaining two added-value products from HTC of sewage sludge, hydrochar as fuel, and P from the remaining wastewater. During HTC, some of P is solubilized, however most can be found in the solid product, due to the presence of Fe, Mg and Ca ions which promote P precipitation as $-PO_4^{3-}$ on the hydrochar surface [48,57,58]. Recently, Becker at al. (2019) developed a method to remove P from hydrochars and precipitate it as struvite [59]. After an acid-leaching step, struvite was precipitated at pH 9 by the ammonium-rich HTC wastewater. At the same time, their hydrochar had a HHV of 13.7 MJ/kg (HTC temperature 220 °C), highlighting that with the necessary fine-tuning, it is possible to produce two high added-value products from sewage sludge. Noticeably, the joint strategy of fuel hydrochars production and P reclamation has also received attention for high-P biomasses, other than sewage sludge [60].

Recently, Aragón-Briceño et al. (2020) achieved P solubilization in the range of 24–27% regardless of initial solids loading (HTC temperature 250 °C, residence time 30 min) [61]. Interestingly, the HHVs of their hydrochars were also rather independent of solid sludge loading, ranging from 15.4–16.5 MJ/kg. Their Aspen Plus analysis indicated a significant positive energy balance when process water and hydrochar were valorized as products. In another study, hydrothermal treatment of sewage sludge digestate at 180–240 °C did not result in high-rank hydrochars, due to the high ash content of the samples. However, when an acid-leaching step was added, lignite-like upgraded hydrochars were obtained and at the same time the acidic leachate was precipitated with the use of CaO to yield a total P content close to 42 mg g$^{-1}$ [62]. The upgrading of hydrochar properties has also been confirmed when organic acids (e.g., oxalic acid) were used to extract and bind P [63]. It is worth noting that in these recent studies, HTC was performed with the natural water content of sewage sludge, without the addition of surplus water. Up to date, the study of Xu et al. (2020) is the only work that has evaluated the effect of aqueous phase recycling on the hydrochar properties, as part of a 'green' environmental engineering approach. The authors determined that the carbon content, nitrogen content and HHV of the hydrochars increased when the aqueous phase was recycled, leading to upgraded hydrochars and improved water use efficiency [64].

As indicated by Citespace, the greenhouse gases (GHGs) emission during the combustion of hydrochars is also an active research topic. The work of Wang et al. (2019) provided a comprehensive insight into the $SO_2$, $NO_x$, and CO emissions during the combustion of sewage sludge hydrochars at 1000 °C. Similarly to earlier works, they obtained their optimum HHV at the HTC temperature of 230 °C [65]. They determined that $SO_2$ emission was inversely correlated to HTC temperature, whereas $NO_x$ emission was almost constant up to the temperature of 260 °C. Although for most feedstocks HTC treatment results in an increased N content (due to the higher losses in organic substance), the opposite trend has been observed in sewage sludge hydrochars [66]. Still, the N content of sewage sludge hydrochars should always be monitored to minimize potential $NO_x$ emissions during the combustion phase. Towards this end, Xu et al. (2020) developed a layered double hydroxide (LDH) catalyst which they mixed with sewage sludge during HTC, in order to minimize the N content of the final product. The catalyst promoted the thermal decomposition of N-organic matter to $NH_4^+$-N, which ended up in the HTC wastewater due to its high aqueous solubility [67].

### 2.3.2. Valorization of Food Waste for the Production of Fuel Hydrochar

Every year there are millions of tons of food waste disposed of in landfills without any form of valorization. As opposed to sewage sludge, food waste is very diverse in nature and heterogeneous in composition and production sources, thus rendering a unified management plan impractical at large-scale. It has been shown that HTC of food waste is a feasible alternative treatment method for the production of high quality fuel hydrochars, since no pretreatment is required and the moisture content is already high [68–70]. It has been generally established that hydrochars with C content and HHVs in the range of 45–93% and 15–30 MJ/kg, respectively, can be obtained from food waste [71–73].

The role and transformation pathway of each basic food component (carbohydrates, proteins, lipids) has provided significant insights. Tradler et al. (2018) collected food waste from restaurants and separated them into vegetal, carbohydrate-rich, and animal-based food. After HTC treatment at 200 °C and 6 hours, they noticed that food high in proteins and fat resulted in lower hydrochar yields than feedstocks high in carbohydrates [74]. The HHV of the homogeneous, mixed food sample was in the order of ~23 MJ/kg. Later, Li et al. (2019) supported these findings and added that the carbohydrate content correlated positively to the fixed carbon content in hydrochar and enhanced the homogenization and thermal stability of the solid biofuel, resulting in a greater combustion performance [75].

Materials developed through pre-designed experimental processes in order to tackle a specific problem or deficiency are called engineered materials. Such an approach was implemented by Akarsu et al. (2019), who combined anaerobic digestion and HTC to convert vegetable and fruit waste to hydrochar with improved fuel properties. Indeed, anaerobic digestion followed by HTC at 250 °C and 30 min treatment time resulted in hydrochar with a HHV of 27.3 MJ/kg and 7.5% ash content [76]. Generally, ash and heavy metals do not appear to be an issue with food waste, although they are typically monitored in most HTC studies. Subsequently, steam gasification of the double-processed waste yielded 33 mol $H_2$/kg of hydrochar at 1050 °C. Nasir et al. (2020) converted spent brewery grains into hydrochars using different solvents during HTC. The group concluded that typical water-based HTC lead to hydrochar with the optimum fuel properties, however the use of methanol, ethanol and 2-propanol resulted in fundamentally different hydrochars, perhaps suitable for soil application or wastewater treatment processes [77]. It is worth noting that such an approach is not often reported in the literature and may be worth investigating further, as part of a multiple-product biorefinery concept.

Defective coffee beans have also been utilized for hydrochar production, with positive results. At 250 °C and 40 min treatment time, the resultant hydrochar had a C content, HHV and ash content of 68.3%, 29.1 MJ/kg, and 0.07%, respectively [78]. As a result, the combustion rate, reactivity and heat release were also improved compared to the respective coffee beans values. However, the low initial moisture content of the beans required the addition of water from an external source. Therefore, it would worth comparing the fuel properties and techno-economic efficiency of these hydrochars to biochars prepared through dry pyrolysis of the beans. Other food waste that have been successfully converted to fuel hydrochars include fruit residues, [79], orange peels [80], and cabbage processing waste [81]. HHVs in the range of 25–30 MJ/kg were reported in these studies.

Similarly to sewage sludge, food waste has been combined with other feedstocks in a co-HTC approach. Wang et al. (2018) combined food waste with wood sawdust to produce hydrochar fuel pellets with improved mechanical and storage characteristics. Hydrochar pellets (HTC 220 °C) with food waste ratios from 50 to 75% exhibited an increased tensile strength, decreased ignition temperature and maximum weight loss rate at a wider temperature range, indicating increased flammability [70].

However, contrary to sewage sludge, it is the food waste being used to upgrade the fuel properties of other materials.

The combination of food waste and low-rank coal appears to be gaining momentum. HHVs up to 31.4 MJ/kg were achieved when blended food waste and coal were mixed on a 1:1 basis and converted to hydrochar. The ash content of hydrochar obtained via the co-HTC at 300 °C was 53% less than the ash content of raw coal [82]. Mazumder et al. (2020a, 2020b) thoroughly investigated the co-HTC of food waste and bituminous coal waste, in an effort to reduce the ash, sulfur, and chloride content of the latter. Their optimum hydrochar (HHV of 23 MJ/kg, sulfur content 1.4%) was obtained at 230 °C and 30 min residence time [83]. Based on these optimum conditions, the authors made further progress by examining the technoeconomic feasibility of the co-HTC process and concluded that the raw material purchasing and transportation cost to be the most influential variable [84]. They noted that the mixture must have about 85% moisture to make sure the positive displacement pumps can pump the feedstock and they proposed to recycle the co-HTC process wastewater to maintain this level of moisture. The recycling of the HTC wastewater to reduce the water requirements of the process has also been supported by others [85].

## 3. Results and Discussion

### 3.1. Statistical Analysis of Hydrochar Properties

The statistical distribution of each one of the 7 hydrochar characteristics produced from biomass by pyrolysis is presented as a boxplot in Figure 2. In order to examine data dispersion, we used the interquartile range (IQR) and divided datasets into quartiles. Looking at each boxplot from top to bottom, five lines representing the maximum, the third quartile (Q3), the median, the first quartile (Q1), and the minimum of the corresponding data can be seen. The rectangular bullet represents the mean value, while the individual circle bullets are the outlier data values, which were not taken into account during the training process. In each plot, the useful data took values between Q1 − 1.5IQR and Q3 + 1.5IQR, where IQR = Q3 − Q1, while the outlier data took values below Q1 − 1.5IQR or above Q3 + 1.5IQR.

With respect to the input parameters, the median (mean) temperature and time during hydrothermal carbonization were 219 °C (227 °C) and 1 hr (3 hr), whereas the range of values was from 150 to 325 °C (Figure 2a), and from 0.01 to 4 hr (Figure 2b), respectively. Moreover, the median (mean) value of the carbon, the oxygen and the hydrogen content of hydrochars was 53% (52%), 28% (29%), and 5.7% (5.6%), while their values varied from 23 to 82% (Figure 2c), from 0.7 to 62% (Figure 2d), and from 3 to 9% (Figure 2e), respectively. On the other hand, for the output parameters, the median (mean) of the higher heating value and the % solid yield was 22 MJ/Kg (22.1 MJ/Kg) and 59% (58%), while their values were in the range of 9–36 MJ/Kg (Figure 2f) and 6–100% (Figure 2g), respectively.

During data acquisition, it was revealed that the variety of original biomass materials used for the production of hydrochars was large (Table 3). In some cases, HTC time was more than 4 h, the carbon content of the resultant hydrochars was less than 23%, and the hydrogen content was less than 3% (Figure 2b,c,e, respectively). Therefore, HTC time, carbon, and hydrogen content present the most outlier values, which could be excluded during neural network training and testing, in order for the ANN models to cover a smaller number of biomasses and achieve better performance. However, it was decided not to sacrifice collected data diversity and be as inclusive as possible in terms of original biomasses as this was one of the strategic objectives of this study. Inevitably, this would cost the accuracy of our ANN models, causing them difficulties in making predictions with absolute accuracy.

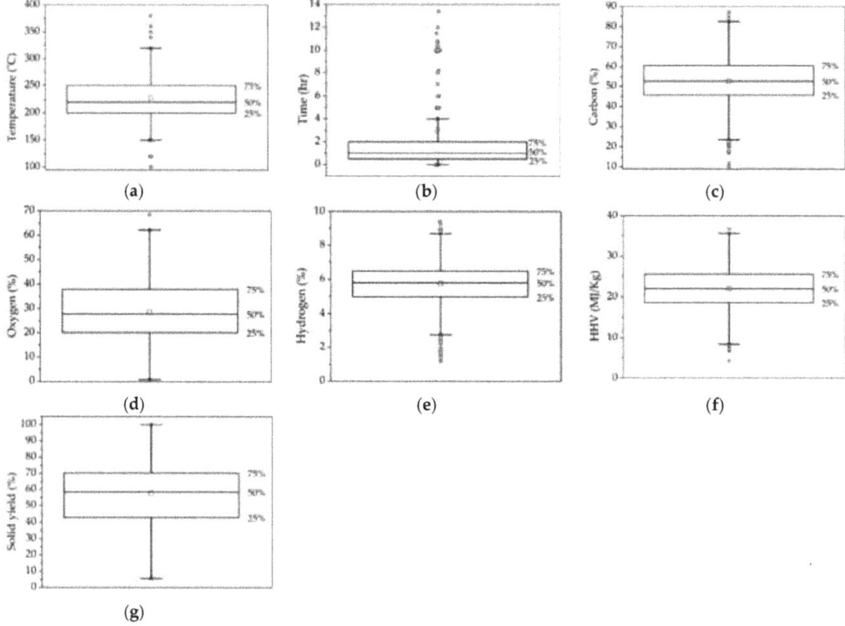

**Figure 2.** Boxplots of the values of (**a**) temperature, (**b**) time, (**c**) carbon, (**d**) oxygen, (**e**) hydrogen, (**f**) HHV, and (**g**) solid yield, related to hydrochar composition.

**Table 3.** Biomasses used for the production of fuel hydrochar and included in data acquisition for the development of the artificial neural networks (ANN) models.

| Biomass | % Appearance in the Related Literature (2014–2020) |
|---|---|
| Sewage sludge | 21.4 |
| Food waste | 15 |
| Corn cob | 12.6 |
| Rice husk | 8.8 |
| Olive mill waste | 7.6 |
| Lower grades of coal | 7.6 |
| Coconut processing residues | 3.8 |
| Miscanthus | 3.8 |
| Banana residues | 3.8 |
| Sugarcane bagasse | <2 |
| Wood sawdust | <2 |
| Paper sludge | <2 |
| Cotton stalk | <2 |
| Eucalyptus leaves | <2 |
| Bamboo residues | <2 |
| Tobacco stalk | <2 |
| Orange peels | <2 |
| Organic fraction of municipal waste | <2 |
| Grape pomace | <2 |
| Poultry litter | <2 |
| Oil palm empty fruit bunch | <2 |

## 3.2. Correlation Patterns between Hydrochar Properties

In order to determine the relationships between any two of the 7 hydrochar characteristics, the Pearson Correlation Coefficient (PCC) $r$ was evaluated, using Equation (1)

$$r_{yz} = \frac{\sum\limits_{i=1}^{N} (y_i - \overline{y}) \sum\limits_{i=1}^{N} (z_i - \overline{z})}{\sqrt{\sum\limits_{i=1}^{N} (y_i - \overline{y})^2} \sqrt{\sum\limits_{i=1}^{N} (z_i - \overline{z})^2}} \tag{1}$$

where, $y$ and $z$ are two randomly selected variables to be examined for linear dependence (correlation), $\overline{y}$ and $\overline{z}$ are their means, and $y_i$ and $z_i$ are individual values of the variables' datasets, respectively.

Table 4 presents the $7 \times 7$ Pearson correlation matrix, revealing correlations, either positive or negative, when the corresponding significance level $p$ is less than 0.01. Therefore, the higher heating value was found to be positively correlated with temperature ($p < 0.01$), carbon content ($p < 0.01$), and hydrogen content ($p < 0.01$), but negatively correlated with oxygen content ($p < 0.01$) and solid yield ($p < 0.01$). Similarly, the carbon content was found to be positively correlated with temperature ($p < 0.01$), hydrogen content ($p < 0.01$), and higher heating value ($p < 0.01$), but negatively correlated with oxygen content ($p < 0.01$) and solid yield ($p < 0.01$).

The relation between any two hydrochar properties was denoted by the value of $r$ from Equation (1), with higher PCC values indicating closer relation. Thus, HHV was closely related to carbon content ($r = 0.886$) and to a lesser extent to oxygen content ($r = -0.411$), hydrogen content ($r = 0.345$), and temperature ($r = 0.321$). The correlation to solid yield ($r = -0.164$) and HTC time ($r = 0.129$) was minimal. Similarly, the carbon content showed a high correlation to HHV, temperature, hydrogen content, and oxygen content, but low correlation to solid yield and time.

In order to investigate these relations and determine the deeper relationship between HHV and its influencing factors, three different ANN models were developed: (1) the $ANN_1$ to predict the HHV values when only the carbon content (the factor with the closer relation) is known, (2) the $ANN_2$ to predict the HHV values when carbon, oxygen, hydrogen, temperature and time (factors with close or loose relation) are given as inputs, and (3) the $ANN_3$ to predict the HHV values when carbon, oxygen, hydrogen, and temperature (only factors with close relation) are given as inputs. Moreover, in order to determine the deep relationship between the carbon content and its influencing factors, we also develop the $ANN_4$ to predict the C content values when HHV, temperature, hydrogen, and oxygen (only factors with close relation) are given as inputs. Keeping in mind the relations suggested by Table 4, we expect that $ANN_3$ will yield the best HHV predictions.

**Table 4.** Pearson correlation matrix between hydrochar properties (CI = 99%).

| PCC | Temperature | Time | Carbon | Hydrogen | Oxygen | HHV | Solid Yield |
|---|---|---|---|---|---|---|---|
| **Temperature** | 1.000 | −0.044 | 0.310 * | −0.158 * | −0.305 * | 0.321 * | −0.299 * |
| **Time** | −0.044 | 1.000 | 0.132 | 0.123 | −0.030 | 0.129 | 0.021 |
| **Carbon** | 0.310 * | 0.132 | 1.000 | 0.286 * | −0.284 * | 0.886 * | −0.182 * |
| **Hydrogen** | −0.158 * | 0.123 | 0.286 * | 1.000 | 0.074 | 0.345 * | −0.107 |
| **Oxygen** | −0.305 * | −0.030 | −0.284 * | 0.074 | 1.000 | −0.411 * | 0.254 * |
| **HHV** | 0.321 * | 0.129 | 0.886 * | 0.345 * | −0.411 * | 1.000 | −0.164 * |
| **Solid Yield** | −0.299 * | 0.021 | −0.182 * | −0.107 | 0.254 * | −0.164 * | 1.000 |

* Denotes significance level $p < 0.01$.

### 3.3. Artificial Neural Network Modeling

The block diagrams of the ANN models applied in this paper are displayed in Figure 3. The multilayer perceptron (MLP) architecture [22,23,86,87], which is popular for similar applications according to the universal approximation theorem, was implemented for each ANN model of Figure 3.

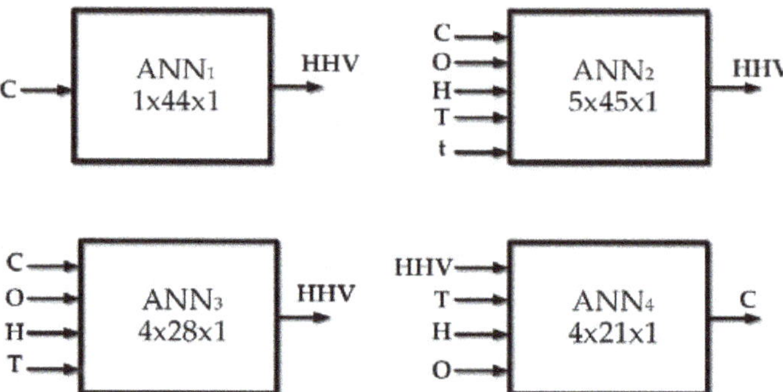


**Figure 3.** ANN models for hydrothermal carbonization. Each ANN box has the input hydrochar parameter(s) on the left, the output parameter on the right, and the number of neurons of the input, the hidden and the output layers inside.

The typical structure of an MLP ANN with three layers is shown in Figure S1 (Supplementary Materials). The number of neurons in each layer is denoted by $M_n$, $n = 1, 2, \ldots, N$. The index $n$ identifies the layers; $n = 1$ refers to the first (input) layer, and $n = N$ to the last (output) layer. The vector $\mathbf{x} = \{x_1, x_2, \ldots, x_{M_0}\}$ denotes the input to ANN's first layer, where $M_0$ is the number of inputs. The vector $\mathbf{y}^n = \{y_1^n, y_2^n, \ldots, y_{M_i}^n\}$ represents the output of the $n$-th layer. The MLP ANN of Figure 3 has a sequential structure, as the output of the $n$-th layer is forwarded to the input of the $(n + 1)$-th layer. Therefore, $\mathbf{y}^n = \Psi^n(\mathbf{y}^{n-1}\mathbf{w}^n + \mathbf{b}^n)$, where the activation function $\Psi^n$ is properly selected, the vector $\mathbf{b}^n$ contains the bias terms of the $n$-th layer, and the matrix $\mathbf{w}^n$ carries the adjustable synaptic weights $w_{i,j}^n$ (with $i = 1, 2, \ldots, M_{n-1}$, $j = 1, 2, \ldots, M_n$), the adjustment of which, is arranged by the appropriate training algorithm.

In order to solve the forward problem of predicting the HHVs of the hydrochars, or the quasi-inverse problem of determining the required C content of hydrochar to achieve a specific HHVs, the ANN models shown in Figure 3 had only one output and one, four or five hydrochar parameters as inputs. For the forward problem, ANN₁, ANN₂, and ANN₃ had 1 output (HHV), and 1 (carbon), 5 (carbon, oxygen, hydrogen, temperature, and time), and 4 (carbon, oxygen, hydrogen, and temperature) inputs, respectively. For the quasi-inverse problem, ANN₄ had 1 output (carbon) and 4 inputs (HHV, temperature, hydrogen, and oxygen).

### 3.4. Data Preprocessing

Before the training procedure, the hydrochar data were normalized into a predefined range, in order to avoid large values that may result in unstable neural networks with poor learning performance and consequently bad generalization. All hydrochar variables were normalized according to Equation (2) into the range $[-1, 1]$:

$$x_i' = \frac{(x_b - x_a)(x_i - x_{min})}{(x_{max} - x_{min})} + x_a \tag{2}$$

where, $x_i'$ is the normalized value of sample $x_i$, $x_b = 1$, $x_a = -1$, and $x_{max}$ and $x_{min}$ are the maximum and minimum values of $x_i$.

After normalization, the data were divided into training, validation and testing distinct datasets, as 70%, 15%, and 15% fractions of the whole data series, respectively. The training dataset was used during the learning process to train and fit the ANN model. The validation dataset was used to validate the ANN model during the hyperparameters' adjustments by the learning algorithm and to early stop the training algorithm in order to avoid overfitting. Finally, the testing dataset, which is an independent set of data kept unseen from the ANN model during training, was used to evaluate the performance and to test the quality of the model [86,87].

The structure of all ANN models of Figure 3 follow the general block diagram of Figure S1 (Supplementary Materials), while the hyperbolic tangent sigmoid activation function was adopted and the Levenberg-Marquardt (LM) learning algorithm was selected in all cases. The learning process was repeated 33 times calculating its average performance, in order to ensure its stability and generalization capability [22]. Finally the number of neurons in the hidden layer was determined when the training Mean Relative Error (MRE), defined by Equation (3), was minimized

$$MRE = \frac{1}{K}\sum_{k=1}^{K}\left(\left|\frac{p_k - e_k}{e_k}\right|\right) \tag{3}$$

where, $e_k$ and $p_k$ stand for the $k$-th experimental and predicted values of the output hydrochar parameters, and $K$ is the multitude of values used for testing. Using the trial and error method, the number of hidden layer neurons was scanned for each of the four ANN models calculating MRE. Its values, varying between 0.04% (best case for $ANN_3$ and $ANN_4$) and 4.3% (worse case for $ANN_2$), are depicted in Figure S2 (Supplementary Materials). Carefully selecting 44, 45, 28, and 21 neurons in the hidden layer of $ANN_1$, $ANN_2$, $ANN_3$, and $ANN_4$, respectively, an error of less than 2.1% was achieved in all cases. This is a satisfactory and well promising outcome for our ANN models.

*3.5. Performance of ANN Models*

When the ANN training and validation processes were completed, the four models implemented herein were tested for their quality and performance, using the testing datasets, which consisted of couples of the form $\left(C_k^{te}, HHV_k^{te}\right)$ for the $ANN_1$, sextets of the form $\left(C_k^{te}, O_k^{te}, H_k^{te}, T_k^{te}, t_k^{te}, HHV_k^{te}\right)$ for the $ANN_2$, quintets of the form $\left(C_k^{te}, O_k^{te}, H_k^{te}, T_k^{te}, HHV_k^{te}\right)$ for the $ANN_3$, and quintets of the form $\left(HHV_k^{te}, T_k^{te}, H_k^{te}, O_k^{te}, C_k^{te}\right)$ for the $ANN_4$, where $k = 1, 2, \ldots, K$ and $C_k^{te}, HHV_k^{te}, O_k^{te}$ and $H_k^{te}$ $T_k^{te}$ and $t_k^{te}$, the testing input values (or required outputs) of the hydrochar C content, HHVs, oxygen and hydrogen content, HTC temperature and duration, respectively. $K$ stands for the number of samples used for testing and has been set equal to 100, 65, 70 or 76, for $ANN_1$, $ANN_2$, $ANN_3$, and $ANN_4$, respectively. The statistical measures calculated herein for the examination of the generalization and prediction ability of the proposed ANNs and the evaluation of the models' performance were the root mean squared error (RMSE) and the regression coefficient ($R^2$), defined as follows:

$$RMSE = \sqrt{\frac{1}{K}\sum_{k=1}^{K}(p_k - e_k)^2} \tag{4}$$

$$R^2 = 1 - \frac{\sum_{k=1}^{K}(p_k - e_k)^2}{\sum_{k=1}^{K}(p_k - \bar{e})^2} \tag{5}$$

where, $e_k$ is the $k$-th experimental value of the output biochar ($HHV_k^{te}$ for the $ANN_1$, $ANN_2$, $ANN_3$ models, or $C_k^{te}$ for the $ANN_4$ model), $\bar{e}$ represents the average of the experimental output values, and

$p_k$ is the $k$-th predicted value of the output biochar ($HHV_k^{pr}$ for the ANN$_1$, ANN$_2$, ANN$_3$ models, or $C_k^{pr}$ for the ANN$_4$ model).

Comparisons of the predicted from the ANN models output hydrochar parameters with their corresponding measured HHVs or C contents are offered in Figure 4. The solid lines correspond to the experimental data deduced from the published literature, whereas the markers represent the predicted values of HHV or the carbon content by the model indicated in the inset. It is apparent that the proposed ANN models were able to predict both HHVs and C contents. Moreover, the performance of ANN$_1$ and ANN$_4$ seemed to be slightly better, especially for values of HHV lower than 20 MJ/Kg and C contents lower than 55%, respectively. Still, the overall performance of all ANNs was of sufficient accuracy.

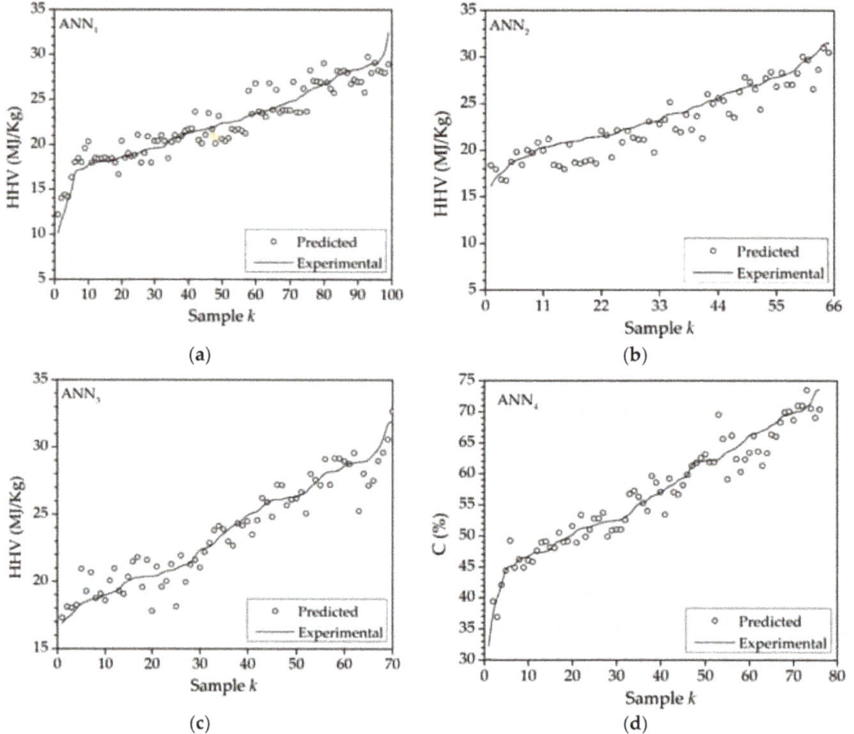


**Figure 4.** Experimental, $HHV_k^{te}$ or $C_k^{te}$, and predicted, $HHV_k^{pr}$ or $C_k^{pr}$ from (**a**) ANN$_1$, (**b**) ANN$_2$, (**c**) ANN$_3$, and (**d**) ANN$_4$ output biochar values.

Finally, the predicted outputs of the ANN models were plotted against their matching experimental values in Figure 5. While the overall RMSE and R$^2$ values developed by our models were acceptable, the ANN$_3$ (with R$^2$ = 0.917 and RMSE = 1.124) appeared to perform better than ANN$_1$ (with R$^2$ = 0.897 and RMSE = 1.289) and ANN$_2$ (with R$^2$ = 0.879 and RMSE = 1.340). The prediction ability of ANN$_3$, in which the RMSE is lower by 13% and 16% than that of ANN$_1$ and ANN$_2$, was superior because HHVs were predicted having as inputs all the closely-correlated hydrochar parameters (C, O, H, and temperature), in contrast to ANN$_1$ and ANN$_2$. Moreover, as shown in Figure 5d, ANN$_4$ achieved a very good performance with a high R$^2$ = 0.943 despite the mediocre RMSE = 2.188.

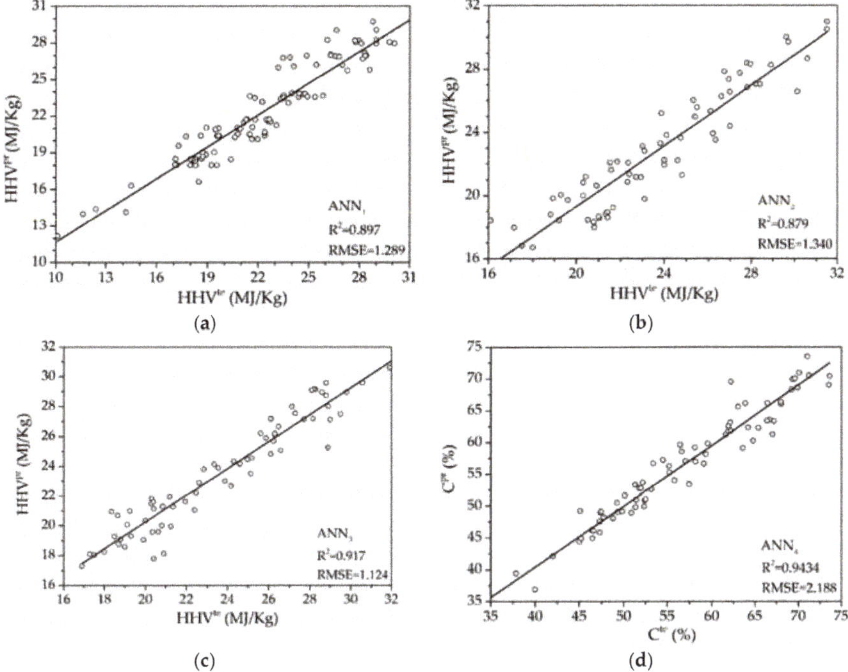

**Figure 5.** Regression analysis of predicted and experimental data (**a**) for higher heating values (HHV) using ANN₁, (**b**) for HHV using ANN₂, (**c**) for HHV using ANN₃, and (**d**) for carbon using ANN₄.

It is evident from Figure 4a–c or Figure 5a–c that our ANN models managed to predict, even though not so accurately in some cases, the HHVs of fuel hydrochars. Minor performance limitations may be attributed to: (1) the heterogeneity of the experimental data, used for training, validation, and testing of our ANN models, as they were gathered from a vast number of literature papers related to a wide range of biomass materials and experimental conditions, (2) the inevitable existence of plethora of multivalued data, as several different published works studying HTC of numerous biomass materials with discrete biochar characteristics (used as input data C, O, H, T, t) resulted in very close or even the same HHVs (used as output data HHV), (3) the non-removal of the outlier collected data values, in order for our ANN models to take into account data from as many biomasses as possible, and (4) the dimensionality of input variables, since model performance is improved only when important input variables are used.

## 4. Conclusions

Conclusively, there is considerable potential in the valorization/co-valorization of various biomasses for the production of hydrochars with improved fuel properties. Sewage sludge and food waste are the main precursor materials at the moment. Processing temperatures and times in the range of 200–230 °C and 30–60 min appear to be the optimum in many cases. Recovery of P through acid-leaching further improves the feasibility of the process and results in hydrochars with lower ash contents. As has always been the case with sewage sludge, monitoring of the fate of heavy metals should always be performed. A significant research gap is the utilization and recycling of various real wastewaters (as feed waters in HTC) on hydrochar properties and subsequent impact on combustion behavior and GHG emission. Water recycling presumes HTC occurs on a dynamic (flowing) mode, a set-up largely unstudied compared to static (batch) conditions. Finally, cost–benefit and life-cycle assessments based on local conditions and feedstocks are missing which are essential before scaling-up of the process. Towards this purpose, the ability of ANNs to predict HHVs of hydrochars regardless of the original biomass used, was demonstrated. Of all the input parameters tested (C, H, O content, HTC temperature and time), C content was found to be the most closely correlated variable to HHV, whereas HTC time showed the least correlation of all. Of the four ANNs developed, $ANN_3$ (based on C, H, O and temperature as inputs) exhibited the optimum performance, however, $ANN_1$ (based only on C content of hydrochars) had a satisfactory performance. Practically, this means that only one laboratory analysis is required for the accurate estimation of the HHV of hydrochars, thus largely reducing the cost and time of research work. In combination with the reverse ANN model ($ANN_4$), researchers will be able to focus on biomasses with a minimum C content for the production of hydrochar, thus adjusting accordingly their HTC temperature (by far the most important process variable). Biomasses with lower initial C content would require higher HTC temperature to improve their fuel properties and resulting in a more energy-demanding process. On the contrary, biomasses with higher initial C content would require milder HTC conditions to achieve hydrochars with the required HHV, thus reducing the cost of the process.

**Supplementary Materials:** The following are available online at http://www.mdpi.com/1996-1073/13/17/4572/s1, Figure S1: Architecture of a feedforward MLP ANN with one input, one hidden and one output layer., Figure S2: MRE (%) as a function of the number of neurons in the hidden layer.

**Author Contributions:** Conceptualization, I.O.V., and D.K.; methodology, T.N.K., C.D.N.; software, I.O.V., T.N.K., and C.D.N.; validation, I.O.V., T.N.K., and C.D.N.; formal analysis, I.O.V., D.K.; investigation, T.K.T., T.T., R.K., A.K.; resources, I.O.V., D.K.; data curation, T.K.T., T.T., R.K., A.K.; writing—original draft preparation, D.K.; writing—review and editing, I.O.V., D.K.; visualization, T.N.K., C.D.N.; supervision, D.K.; project administration, D.K.; All authors have read and agreed to the published version of the manuscript.

**Funding:** This research received no external funding.

**Conflicts of Interest:** The authors declare no conflict of interest.

## References

1. Jiménez-Carmona, M.M.; Luque de Castro, M.D. Isolation of eucalyptus essential oil for GC-MS analysis by extraction with subcritical water. *Chromatographia* **1999**, *50*, 578–582. [CrossRef]

2. Jiménez-Carmona, M.M.; Ubera, J.L.; Luque De Castro, M.D. Comparison of continuous subcritical water extraction and hydrodistillation of marjoram essential oil. *J. Chromatogr. A* **1999**, *855*, 625–632. [CrossRef]

3. Rovio, S.; Hartonen, K.; Holm, Y.; Hiltunen, R.; Riekkola, M.L. Extraction of clove using pressurized hot water. *Flavour Fragr. J.* **1999**, *14*, 399–404. [CrossRef]

4. Yang, Y.; Jones, A.D.; Eaton, C.D. Retention behavior of phenols, anilines, and alkylbenzenes in liquid chromatographic separations using subcritical water as the mobile phase. *Anal. Chem.* **1999**, *71*, 3808–3813. [CrossRef]

5. Yang, Y.; Li, B. Subcritical water extraction coupled to high-performance liquid chromatography. *Anal. Chem.* **1999**, *71*, 1491–1495. [CrossRef]

6. Hawthorne, S.B.; Lagadec, A.J.M.; Kalderis, D.; Lilke, A.V.; Miller, D.J. Pilot-scale destruction of TNT, RDX, and HMX on contaminated soils using subcritical water. *Environ. Sci. Technol.* **2000**, *34*, 3224–3228. [CrossRef]

7. Kubatova, A.; Lagadec, A.J.M.; Hawthorne, S.B. Dechlorination of lindane, dieldrin, tetrachloroethane, trichloroethene and PVC in Subcritical Water. *Environ. Sci. Technol.* **2002**, *36*, 1337–1343. [CrossRef]

8. Daskalaki, V.M.; Timotheatou, E.S.; Katsaounis, A.; Kalderis, D. Degradation of Reactive Red 120 using hydrogen peroxide in subcritical water. *Desalination* **2011**, *274*, 200–205. [CrossRef]

9. Kirmizakis, P.; Tsamoutsoglou, C.; Kayan, B.; Kalderis, D. Subcritical water treatment of landfill leachate: Application of response surface methodology. *J. Environ. Manag.* **2014**, *146*, 9–15. [CrossRef]

10. Skubiszewska-Zi, J.; Charmas, B.; Leboda, R.; Staszczuk, P.; Kowalczyk, P.; Oleszczuk, P. Effect of hydrothermal modification on the porous structure and thermal properties of carbon-silica adsorbents (carbosils). *Mater. Chem. Phys.* **2003**, *78*, 486–494. [CrossRef]

11. Yu, S.H.; Cui, X.; Li, L.; Li, K.; Yu, B.; Antonietti, M.; Cölfen, H. From starch to metal/carbon hybrid nanostructures: Hydrothermal metal-catalyzed carbonization. *Adv. Mater.* **2004**, *16*, 1636–1640. [CrossRef]

12. Sarkar, N.B.; Sarkar, P.; Choudhury, A. Effect of hydrothermal treatment of coal on the oxidation susceptibility and electrical resistivity of HTT coke. *Fuel Process. Technol.* **2005**, *86*, 487–497. [CrossRef]
13. Funke, A.; Ziegler, F. Hydrothermal carbonization of biomass: A summary and discussion of chemical mechanisms for process engineering. *Biofuel Bioprod. Biorefin.* **2010**, *4*, 160–177. [CrossRef]
14. Libra, J.A.; Ro, K.S.; Kammann, C.; Funke, A.; Berge, N.D.; Neubauer, Y.; Titirici, M.-M.; Führer, C.; Bens, O.; Kern, J.; et al. Hydrothermal carbonization of biomass residuals: A comparative review of the chemistry, processes and applications of wet and dry pyrolysis. *Biofuels* **2011**, *2*, 71–106. [CrossRef]
15. Heidari, M.; Dutta, A.; Acharya, B.; Mahmud, S. A review of the current knowledge and challenges of hydrothermal carbonization for biomass conversion. *J. Energy Inst.* **2019**, *92*, 1779–1799. [CrossRef]
16. Wang, T.; Zhai, Y.; Zhu, Y.; Li, C.; Zeng, G. A review of the hydrothermal carbonization of biomass waste for hydrochar formation: Process conditions, fundamentals, and physicochemical properties. *Renew. Sustain. Energy Rev.* **2018**, *90*, 223–247. [CrossRef]
17. McGaughy, K.; Toufiq Reza, M. Hydrothermal carbonization of food waste: Simplified process simulation model based on experimental results. *Biomass Convers. Biorefin.* **2018**, *8*, 283–292. [CrossRef]
18. Gallifuoco, A. A new approach to kinetic modeling of biomass hydrothermal carbonization. *ACS Sustain. Chem. Eng.* **2019**, *7*, 13073–13080. [CrossRef]
19. Conag, A.T.; Villahermosa, J.E.R.; Cabatingan, L.K.; Go, A.W. Predictive HHV model for raw and torrefied sugarcane residues. *Waste Biomass Valorization* **2019**, *10*, 1929–1943. [CrossRef]
20. Vallejo, F.; Díaz-Robles, L.A.; Vega, R.; Cubillos, F. A novel approach for prediction of mass yield and higher calorific value of hydrothermal carbonization by a robust multilinear model and regression trees. *J. Energy Inst.* **2020**, *93*, 1755–1762. [CrossRef]
21. Akdeniz, F.; Biçil, M.; Karadede, Y.; Özbek, F.E.; Özdemir, G. Application of real valued genetic algorithm on prediction of higher heating values of various lignocellulosic materials using lignin and extractive contents. *Energy* **2018**, *160*, 1047–1054. [CrossRef]
22. Kapetanakis, T.N.; Vardiambasis, I.O.; Ioannidou, M.P.; Maras, A. Neural network modeling for the solution of the inverse loop antenna radiation problem. *IEEE Trans. Antennas Propag.* **2018**, *66*, 6283–6290. [CrossRef]
23. Sergaki, E.; Spiliotis, G.; Vardiambasis, I.O.; Kapetanakis, T.; Krasoudakis, A.; Giakos, G.C.; Zervakis, M.; Polydorou, A. Application of ANN and ANFIS for Detection of Brain Tumors in MRIs by Using DWT and GLCM Texture Analysis. In Proceedings of the IST 2018—International Conference on Imaging Systems and Techniques, Krakow, Poland, 16–18 October 2018; pp. 1–6.
24. Bhange, V.P.; Bhivgade, U.V.; Vaidya, A.N. Artificial Neural Network Modeling in Pretreatment of Garden Biomass for Lignocellulose Degradation. *Waste Biomass Valorization* **2019**, *10*, 1571–1583. [CrossRef]
25. Baruah, D.; Baruah, D.C.; Hazarika, M.K. Artificial neural network based modeling of biomass gasification in fixed bed downdraft gasifiers. *Biomass Bioenergy* **2017**, *98*, 264–271. [CrossRef]
26. Nasrudin, N.A.; Jewaratnam, J.; Hossain, M.A.; Ganeson, P.B. Performance comparison of feedforward neural network training algorithms in modelling microwave pyrolysis of oil palm fibre for hydrogen and biochar production. *Asia-Pac. J. Chem. Eng.* **2020**, *15*. [CrossRef]
27. Chen, C.; Liu, G.; An, Q.; Lin, L.; Shang, Y.; Wan, C. From wasted sludge to valuable biochar by low temperature hydrothermal carbonization treatment: Insight into the surface characteristics. *J. Clean. Prod.* **2020**, *263*, 121600. [CrossRef]
28. Zheng, X.; Jiang, Z.; Ying, Z.; Song, J.; Chen, W.; Wang, B. Role of feedstock properties and hydrothermal carbonization conditions on fuel properties of sewage sludge-derived hydrochar using multiple linear regression technique. *Fuel* **2020**, *271*, 117609. [CrossRef]
29. Chen, C.; Ibekwe-Sanjuan, F.; Hou, J. The Structure and Dynamics of Co-Citation Clusters: A Multiple-Perspective the Structure and Dynamics of Co-Citation Clusters: A Multiple-Perspective Co-Citation Analysis. *J. Am. Soc. Inf. Sci. Technol.* **2010**, *61*, 1386–1409. [CrossRef]
30. Chen, C. CiteSpace II: Detecting and visualizing emerging trends and transient patterns in scientific literature. *J. Assoc. Inf. Sci. Technol.* **2006**, *57*, 359–377. [CrossRef]
31. Fang, Y.; Yin, J.; Wu, B. Climate change and tourism: A scientometric analysis using CiteSpace. *J. Sustain. Tour.* **2018**, *26*, 108–126. [CrossRef]
32. Wu, P.; Wang, Z.; Wang, H.; Bolan, N.S.; Wang, Y.; Chen, W. Visualizing the emerging trends of biochar research and applications in 2019: A scientometric analysis and review. *Biochar* **2020**, *2*, 135–150. [CrossRef]

33. Teoh, S.K.; Li, L.Y. Feasibility of alternative sewage sludge treatment methods from a lifecycle assessment (LCA) perspective. *J. Clean. Prod.* **2020**, *247*, 119495. [CrossRef]

34. Cieślik, B.M.; Namieśnik, J.; Konieczka, P. Review of sewage sludge management: Standards, regulations and analytical methods. *J. Clean. Prod.* **2015**, *90*, 1–15. [CrossRef]

35. Chang, Z.; Long, G.; Zhou, J.L.; Ma, C. Valorization of sewage sludge in the fabrication of construction and building materials: A review. *Resour. Conserv. Recycl.* **2020**, *154*, 104606. [CrossRef]

36. Zhao, P.; Shen, Y.; Ge, S.; Yoshikawa, K. Energy recycling from sewage sludge by producing solid biofuel with hydrothermal carbonization. *Energy Convers. Manag.* **2014**, *78*, 815–821. [CrossRef]

37. Kim, D.; Lee, K.; Park, K.Y. Hydrothermal carbonization of anaerobically digested sludge for solid fuel production and energy recovery. *Fuel* **2014**, *130*, 120–125. [CrossRef]

38. Parshetti, G.K.; Liu, Z.; Jain, A.; Srinivasan, M.P.; Balasubramanian, R. Hydrothermal carbonization of sewage sludge for energy production with coal. *Fuel* **2013**, *111*, 201–210. [CrossRef]

39. Gai, C.; Chen, M.; Liu, T.; Peng, N.; Liu, Z. Gasification characteristics of hydrochar and pyrochar derived from sewage sludge. *Energy* **2016**, *113*, 957–965. [CrossRef]

40. Danso-Boateng, E.; Shama, G.; Wheatley, A.D.; Martin, S.J.; Holdich, R.G. Hydrothermal carbonisation of sewage sludge: Effect of process conditions on product characteristics and methane production. *Bioresour. Technol.* **2015**, *177*, 318–327. [CrossRef]

41. Silva, R.D.V.K.; Lei, Z.; Shimizu, K.; Zhang, Z. Hydrothermal treatment of sewage sludge to produce solid biofuel: Focus on fuel characteristics. *Bioresour. Technol. Rep.* **2020**, *11*, 100453. [CrossRef]

42. Mäkelä, M.; Fullana, A.; Yoshikawa, K. Ash behavior during hydrothermal treatment for solid fuel applications. Part 1: Overview of different feedstock. *Energy Convers. Manag.* **2016**, *121*, 402–408. [CrossRef]

43. Mäkelä, M.; Yoshikawa, K. Ash behavior during hydrothermal treatment for solid fuel applications. Part 2: Effects of treatment conditions on industrial waste biomass. *Energy Convers. Manag.* **2016**, *121*, 409–414. [CrossRef]

44. Smith, A.M.; Singh, S.; Ross, A.B. Fate of inorganic material during hydrothermal carbonisation of biomass: Influence of feedstock on combustion behaviour of hydrochar. *Fuel* **2016**, *169*, 135–145. [CrossRef]

45. Parmar, K.R.; Ross, A.B. Integration of hydrothermal carbonisation with anaerobic digestion; Opportunities for valorisation of digestate. *Energies* **2019**, *12*, 1586. [CrossRef]

46. Wang, L.; Chang, Y.; Li, A. Hydrothermal carbonization for energy-efficient processing of sewage sludge: A review. *Renew. Sustain. Energy Rev.* **2019**, *108*, 423–440. [CrossRef]

47. Tasca, A.L.; Puccini, M.; Gori, R.; Corsi, I.; Galletti, A.M.R.; Vitolo, S. Hydrothermal carbonization of sewage sludge: A critical analysis of process severity, hydrochar properties and environmental implications. *Waste Manag.* **2019**, *93*, 1–13. [CrossRef]

48. Niinipuu, M.; Latham, K.G.; Boily, J.F.; Bergknut, M.; Jansson, S. The impact of hydrothermal carbonization on the surface functionalities of wet waste materials for water treatment applications. *Environ. Sci. Pollut. Res.* **2020**, *27*, 24369–24379. [CrossRef]

49. Ma, J.; Chen, M.; Yang, T.; Liu, Z.; Jiao, W.; Li, D.; Gai, C. Gasification performance of the hydrochar derived from co-hydrothermal carbonization of sewage sludge and sawdust. *Energy* **2019**, *173*, 732–739. [CrossRef]

50. Ma, J.; Luo, H.; Li, Y.; Liu, Z.; Li, D.; Gai, C.; Jiao, W. Pyrolysis kinetics and thermodynamic parameters of the hydrochars derived from co-hydrothermal carbonization of sawdust and sewage sludge using thermogravimetric analysis. *Bioresour. Technol.* **2019**, *282*, 133–141. [CrossRef]

51. Song, Y.; Zhan, H.; Zhuang, X.; Yin, X.; Wu, C. Synergistic characteristics and capabilities of co-hydrothermal carbonization of sewage sludge/lignite mixtures. *Energy Fuels* **2019**, *33*, 8735–8745. [CrossRef]

52. Wang, L.; Chang, Y.; Zhang, X.; Yang, F.; Li, Y.; Yang, X.; Dong, S. Hydrothermal co-carbonization of sewage sludge and high concentration phenolic wastewater for production of solid biofuel with increased calorific value. *J. Clean. Prod.* **2020**, *255*, 120317. [CrossRef]

53. Lee, J.; Sohn, D.; Lee, K.; Park, K.Y. Solid fuel production through hydrothermal carbonization of sewage sludge and microalgae *Chlorella* sp. from wastewater treatment plant. *Chemosphere* **2019**, *230*, 157–163. [CrossRef] [PubMed]

54. Xu, Z.X.; Song, H.; Zhang, S.; Tong, S.Q.; He, Z.X.; Wang, Q.; Li, B.; Hu, X. Co-hydrothermal carbonization of digested sewage sludge and cow dung biogas residue: Investigation of the reaction characteristics. *Energy* **2019**, *187*, 115972. [CrossRef]

55. Zheng, C.; Ma, X.; Yao, Z.; Chen, X. The properties and combustion behaviors of hydrochars derived from co-hydrothermal carbonization of sewage sludge and food waste. *Bioresour. Technol.* **2019**, *285*, 121347. [CrossRef]

56. He, C.; Zhang, Z.; Ge, C.; Liu, W.; Tang, Y.; Zhuang, X.; Qiu, R. Synergistic effect of hydrothermal co-carbonization of sewage sludge with fruit and agricultural wastes on hydrochar fuel quality and combustion behavior. *Waste Manag.* **2019**, *100*, 171–181. [CrossRef]

57. Heilmann, S.M.; Molde, J.S.; Timler, J.G.; Wood, B.M.; Mikula, A.L.; Vozhdayev, G.V.; Colosky, E.C.; Spokas, K.A.; Valentas, K.J. Phosphorus reclamation through hydrothermal carbonization of animal manures. *Environ. Sci. Technol.* **2014**, *48*, 10323–10329. [CrossRef]

58. Ovsyannikova, E.; Arauzo, P.J.; Becker, G.; Kruse, A. Experimental and thermodynamic studies of phosphate behavior during the hydrothermal carbonization of sewage sludge. *Sci. Total Environ.* **2019**, *692*, 147–156. [CrossRef]

59. Becker, G.C.; Wüst, D.; Köhler, H.; Lautenbach, A.; Kruse, A. Novel approach of phosphate-reclamation as struvite from sewage sludge by utilising hydrothermal carbonization. *J. Environ. Manag.* **2019**, *238*, 119–125. [CrossRef]

60. Cui, X.; Lu, M.; Khan, M.B.; Lai, C.; Yang, X.; He, Z.; Chen, G.; Yan, B. Hydrothermal carbonization of different wetland biomass wastes: Phosphorus reclamation and hydrochar production. *Waste Manag.* **2020**, *102*, 106–113. [CrossRef]

61. Aragón-Briceño, C.I.; Grasham, O.; Ross, A.B.; Dupont, V.; Camargo-Valero, M.A. Hydrothermal carbonization of sewage digestate at wastewater treatment works: Influence of solid loading on characteristics of hydrochar, process water and plant energetics. *Renew. Energy* **2020**, *157*, 959–973. [CrossRef]

62. Marin-Batista, J.D.; Mohedano, A.F.; Rodríguez, J.J.; de la Rubia, M.A. Energy and phosphorous recovery through hydrothermal carbonization of digested sewage sludge. *Waste Manag.* **2020**, *105*, 566–574. [CrossRef]

63. Song, E.; Park, S.; Kim, H. Upgrading hydrothermal carbonization (HTC) hydrochar from sewage sludge. *Energies* **2019**, *12*, 2383. [CrossRef]

64. Xu, Z.X.; Song, H.; Li, P.J.; He, Z.X.; Wang, Q.; Wang, K.; Duan, P.G. Hydrothermal carbonization of sewage sludge: Effect of aqueous phase recycling. *Chem. Eng. J.* **2020**, *387*, 123410. [CrossRef]

65. Wang, R.; Wang, C.; Zhao, Z.; Jia, J.; Jin, Q. Energy recovery from high-ash municipal sewage sludge by hydrothermal carbonization: Fuel characteristics of biosolid products. *Energy* **2019**, *186*, 115848. [CrossRef]

66. Hansen, L.J.; Fendt, S.; Spliethoff, H. Impact of hydrothermal carbonization on combustion properties of residual biomass. *Biomass Convers. Biorefin.* **2020**. [CrossRef]

67. Xu, Z.X.; Song, H.; Li, P.J.; Zhu, X.; Zhang, S.; Wang, Q.; Duan, P.G.; Hu, X. A new method for removal of nitrogen in sewage sludge-derived hydrochar with hydrotalcite as the catalyst. *J. Hazard. Mater.* **2020**, *398*, 122833. [CrossRef]

68. Lin, Y.; Ma, X.; Peng, X.; Yu, Z. Hydrothermal carbonization of typical components of municipal solid waste for deriving hydrochars and their combustion behavior. *Bioresour. Technol.* **2017**, *243*, 539–547. [CrossRef]

69. Gallifuoco, A.; Taglieri, L.; Scimia, F.; Papa, A.A.; Di Giacomo, G. Hydrothermal carbonization of Biomass: New experimental procedures for improving the industrial Processes. *Bioresour. Technol.* **2017**, *244*, 160–165. [CrossRef]

70. Wang, T.; Zhai, Y.; Li, H.; Zhu, Y.; Li, S.; Peng, C.; Wang, B.; Wang, Z.; Xi, Y.; Wang, S.; et al. Co-hydrothermal carbonization of food waste-woody biomass blend towards biofuel pellets production. *Bioresour. Technol.* **2018**, *267*, 371–377. [CrossRef]

71. Saqib, N.U.; Sharma, H.B.; Baroutian, S.; Dubey, B.; Sarmah, A.K. Valorisation of food waste via hydrothermal carbonisation and techno-economic feasibility assessment. *Sci. Total Environ.* **2019**, *690*, 261–276. [CrossRef]

72. Zhao, K.; Li, Y.; Zhou, Y.; Guo, W.; Jiang, H.; Xu, Q. Characterization of hydrothermal carbonization products (hydrochars and spent liquor) and their biomethane production performance. *Bioresour. Technol.* **2018**, *267*, 9–16. [CrossRef] [PubMed]

73. Saqib, N.U.; Baroutian, S.; Sarmah, A.K. Physicochemical, structural and combustion characterization of food waste hydrochar obtained by hydrothermal carbonization. *Bioresour. Technol.* **2018**, *266*, 357–363. [CrossRef] [PubMed]

74. Tradler, S.B.; Mayr, S.; Himmelsbach, M.; Priewasser, R.; Baumgartner, W.; Stadler, A.T. Hydrothermal carbonization as an all-inclusive process for food-waste conversion. *Bioresour. Technol. Rep.* **2018**, *2*, 77–83. [CrossRef]

75. Li, Y.; Liu, H.; Xiao, K.; Jin, M.; Xiao, H.; Yao, H. Combustion and Pyrolysis Characteristics of Hydrochar Prepared by Hydrothermal Carbonization of Typical Food Waste: Influence of Carbohydrates, Proteins, and Lipids. *Energy Fuels* **2020**, *34*, 430–439. [CrossRef]
76. Akarsu, K.; Duman, G.; Yilmazer, A.; Keskin, T.; Azbar, N.; Yanik, J. Sustainable valorization of food wastes into solid fuel by hydrothermal carbonization. *Bioresour. Technol.* **2019**, *292*, 121959. [CrossRef]
77. Atiqah Nasir, N.; Davies, G.; McGregor, J. Tailoring product characteristics in the carbonisation of brewers' spent grain through solvent selection. *Food Bioprod. Process.* **2020**, *120*, 41–47. [CrossRef]
78. Santos Santana, M.; Pereira Alves, R.; da Silva Borges, W.M.; Francisquini, E.; Guerreiro, M.C. Hydrochar production from defective coffee beans by hydrothermal carbonization. *Bioresour. Technol.* **2020**, *300*, 122653. [CrossRef]
79. Zhang, B.; Heidari, M.; Regmi, B.; Salaudeen, S.; Arku, P.; Thimmannagari, M.; Dutta, A. Hydrothermal carbonization of fruit wastes: A promising technique for generating hydrochar. *Energies* **2018**, *11*, 2022. [CrossRef]
80. Xiao, K.; Liu, H.; Li, Y.; Yi, L.; Zhang, X.; Hu, H.; Yao, H. Correlations between hydrochar properties and chemical constitution of orange peel waste during hydrothermal carbonization. *Bioresour. Technol.* **2018**, *265*, 432–436. [CrossRef]
81. Tamelová, B.; Maľaťák, J.; Velebil, J. Hydrothermal carbonization and torrefaction of cabbage waste. *Agron. Res.* **2019**, *17*, 862–871.
82. Ul Saqib, N.; Sarmah, A.K.; Baroutian, S. Effect of temperature on the fuel properties of food waste and coal blend treated under co-hydrothermal carbonization. *Waste Manag.* **2019**, *89*, 236–246. [CrossRef] [PubMed]
83. Mazumder, S.; Saha, P.; Reza, M.T. Co-hydrothermal carbonization of coal waste and food waste: Fuel characteristics. *Biomass Convers. Biorefin.* **2020**. [CrossRef]
84. Mazumder, S.; Saha, P.; McGaughy, K.; Saba, A.; Reza, M.T. Technoeconomic analysis of co-hydrothermal carbonization of coal waste and food waste. *Biomass Convers. Biorefin.* **2020**. [CrossRef]
85. Gupta, D.; Mahajani, S.M.; Garg, A. Effect of hydrothermal carbonization as pretreatment on energy recovery from food and paper wastes. *Bioresour. Technol.* **2019**, *285*, 121329. [CrossRef] [PubMed]
86. Kapetanakis, T.N.; Vardiambasis, I.O.; Lourakis, E.I.; Maras, A. Applying neuro-fuzzy soft computing techniques to the circular loop antenna radiation problem. *IEEE Antennas Wirel. Propag. Lett.* **2018**, *17*, 1673–1676. [CrossRef]
87. Liodakis, G.; Arvanitis, D.; Vardiambasis, I.O. Neural network–based digital receiver for radio communications. *WSEAS Trans. Syst.* **2004**, *3*, 3308–3313.

*Article*

# Assessment of Agro-Environmental Impacts for Supplemented Methods to Biochar Manure Pellets during Rice (*Oryza sativa* L.) Cultivation

JoungDu Shin [1,*], SangWon Park [2] and Changyoon Jeong [3]

[1] Department of Climate Change and Agro-ecology, National Institute of Agricultural Sciences, WanJu Gun 55365, Korea

[2] Chemical Safety Division, National Institute of Agricultural Sciences, WanJu Gun 55365, Korea; swpark@rda.go.kr

[3] Red River Research Station Louisiana State University AgCenter, 262 Research Station Driver Bossier City, Louisiana, LA 7112, USA; CJeong@agcenter.lsu.edu

* Correspondence: jdshin1@korea.kr

Received: 19 March 2020; Accepted: 14 April 2020; Published: 21 April 2020

**Abstract:** The agro-environmental impact of supplemented biochar manure pellet fertilizer (SBMPF) application was evaluated by exploring changes of the chemical properties of paddy water and soil, carbon sequestration, and grain yield during rice cultivation. The treatments consisted of (1) the control (no biochar), (2) pig manure compost pellet (PMCP), (3) biochar manure pellets (BMP) with urea solution heated at 60 °C (BMP-U60), (4) BMP with N, P, and K solutions at room temperature (BMP-NPK), and (5) BMP with urea and K solutions at room temperature (BMP-UK). The $NO_3^--N$ and $PO_4^--P$ concentrations in the control and PMCP in the paddy water were relatively higher compared to SBMPF applied plots. For paddy soil, $NH_4^+-N$ concentration in the control was lower compared to the other SBMPFs treatments 41 days after rice transplant. Additionally, it is possible that the SBMPFs could decrease the phosphorus levels in agricultural ecosystems. Also, the highest carbon sequestration was 2.67 tonnes C ha$^{-1}$ in the BMP-UK treatment, while the lowest was 1.14 tonnes C ha$^{-1}$ in the BMP-U60 treatment. The grain yields from the SBMPFs treatments except for the BMP-UK were significantly higher than the control. Overall, it appeared that the supplemented BMP-NPK application was one of the best SBMPFs considered with respect to agro-environmental impacts during rice cultivation.

**Keywords:** Mitigation of $CO_2$-equiv.; nutrient release; rice paddy water and soil system; slow-release fertilizer

## 1. Introduction

Developing methodologies to improve crop productivity and protect soil systems while mitigating environmental pollution is the current direction of research in sustainable agriculture [1–3]. Recently, biomass conversion from agricultural wastes to carbon-rich materials such as biochar has been recognized as a promising option to maintain or increase soil productivity [4], reduce nutrient losses [5], and mitigate greenhouse gas emissions [6] from the agroecosystem. It is estimated that 50 million tonnes of the 80 million tonnes of organic wastes produced in Korea originate from agriculture [7]. Carbon sequestration utilizing recycled organic wastes through biomass conservation technology can greatly mitigate greenhouse gas emissions and the environmental impact of organic waste in Korea. Biochar is made through the pyrolysis under high temperature in oxygen-limited conditions [8]. Converted biochar from agricultural biomass becomes recalcitrant carbonaceous structures. The structures and components of biochars are strongly related to the source of feedstock and the operating conditions that are used in biochar production. Cantrell et al. [9] documented that the biochar made of poultry litter presented a relatively high nutrient content comparable to fertilizer. The reported analytical characterization of biochar is ranges between 5.2–10.3 in pH, 1.1–55.8% in ash content, 23.6–87.5% in carbon content, and 0–642 $m^2$ $g^{-1}$ in surface area [8,10,11]. Kim et al. [12] reported ranges of 10–69 $cmol_c$ $kg^{-1}$ in the cation exchange capacity (CEC) of biochar. Biochar application can significantly increase plant growth, crop yield, and root biomass by enhancing nutrient use efficiency [13,14]. However, few studies have reported a negative growth response in the early stages of plant growth [15,16]. Thus, research on the incorporation of biochar as a soil amendment in crop fields is still required to improve the production methods and application of biochar in soil. Drift of biochar occurs during field application due to the low density and irregular particle size of biochar. Husk and Major [17] reported that the biochar drift during field application was 25%, while the surface runoff losses due to intense rain events were estimated from 20% to 53% of incorporated biochar [18]. Pelletizing biochar can be a possible solution to minimize losses during field application, and it can also reduce handling and transportation costs [19].

Animal waste composts are recognized as valuable sources of major plant nutrients that reduce the need for synthetic fertilizers [20]. However, environmental problems such as nutrient loss due to surface runoff may arise if excess manure is applied to the agricultural land in sensitive catchment areas. One of the critical issues plaguing animal waste compost application is the lack of an environmentally safe application method to agricultural land in order to mitigate non-point source pollution [21,22]. Most of the nutrients losses from agricultural lands are caused by soil erosion from irrigated agriculture or runoff and leaching after rainfall events [23]. Hence, the top priority was to develop methods that would minimize rapid nutrient loss from animal waste manure application and mitigate nutrient runoff after irrigation or rainfall events. Major pathways of N losses are $NH_4{}^+$–N and $NO_3{}^-$–N leaching, $NH_3$ volatilization, and runoff losses. New strategies such as biochar-manure pelletizing methods are available to minimize N loss from the application of animal-waste compost. New approaches that would improve the efficiency of compost are significant to agricultural production in Korea, because the amount of animal waste must be disposed in an effective manner with a minimal impact on agricultural eco-systems.

In general, the production of biochar pellets with poultry litter mixed with switch grass (BMP) is relatively simple. Pellet is blended poultry litter with powder of switchgrass, and then BMP is produced with slow pyrolysis [24]. Several scientists reported that the synergistic effects of biochar blended with inorganic fertilizer or biochar mixed with nutrient-rich compost were observed to improve crop yields [25–27]. There is only limited information on the field application of supplemented biochar manure pellets with inorganic fertilizers (SBMPFs). SBMPF provides supplemental nutrients and can also regulate nutrient loss or release rate by functioning as a slow release fertilizer. Slow-release fertilizers gradually discharge nutrients to the soil during the growing season and provide sufficient nutrients to crops while minimizing leaching losses [28], which can increase farmers' profits and minimize environmental impacts [29]. Ultimately, this application ameliorates the loss of income in agro-business and mitigates the potential contamination of agricultural watersheds. SBMPFs thus represent an efficient way to decrease field application costs and biochar loss during soil application [19].

However, only limited information on blended biochar pellets functioning as slow-release fertilizers is available. Kim et al. [30] indicated that the application of a combination of biochar and slow release fertilizers yielded the lowest methane emissions among the treatments due to the inhibition of methanogenic bacteria via increased soil aeration and improved rice yield compared to the control.

Additional benefit for cropland application of biochar is carbon sequestration [31,32]. Biochar has a much longer residency period (up to 1000 years) compared to raw materials because of its recalcitrance to biotic and abiotic degradation [33]. However, biochar is partly degraded and oxidized into $CO_2$ when incorporated into soils [34] and up to 50% of feedstock carbon may be lost during pyrolysis [31,35]. Therefore, reduction of carbon during biochar production and increasing its stability in the soil would improve its potential for carbon sequestration. In terms of soil carbon sequestration and the mitigation of $CO_2$-equiv. (carbon dioxide equivalency) emission, biochar incorporated with cow manure compost can sequester 2.3 tonnes C ha$^{-1}$, and ranges from 7.3 to 8.4 tonnes ha$^{-1}$ for mitigating $CO_2$-equiv. emission in the cornfield [36]. Shin et al. [37] indicated that the application of biochar pellets blended with organic compost is a promising way to increase carbon sequestration during crop cultivation. For the application of BMP, carbon sequestration and mitigation of $CO_2$-equiv. emission were 1.65 tonnes ha$^{-1}$ and 6.06 tonnes ha$^{-1}$ greater than those of the control, respectively, during rice cultivation [38]. Soil carbon sequestration from the application of biochar made of wood branch increased from 1.87 to 13.37 tonnes ha$^{-1}$, while the plots with rice straw application demonstrated decreased soil carbon from 2.56 to 0.92 tonnes ha$^{-1}$ [39].

The objective of this study was to evaluate the agro-environmental impact of supplemented biochar manure pellet fertilizers (SBMPFs) application on the agro-ecosystems and soil carbon sequestration during the rice growing season. It is hypothesized that the SBMPFs can significantly mitigate non-point pollution sources and increase potential carbon sequestration in agro-ecosystems.

## 2. Materials and Methods

### 2.1. Biochar Production

Biochar derived from rice hull was purchased from a local farming cooperative society in Go-chang, JeonBuk, South Korea. The top to bottom pyrolysis method to produce biochar was employed, wherein rice hull is burned from the upper level to bottom, and reduces oxygen flux from the exterior of the pyrolysis system at 29.4 KPa of air suction rate. The maximum temperatures during pyrolysis were from 490 °C at the top and 550 °C at the bottom of the pyrolysis system. The loading volume in each batch was 1.5 m$^3$ of rice. The biochar was milled with a grinder to pass through a 2-mm sieve before chemical analysis. The same raw materials were used for both the biochar and pig manure compost, and their chemical properties are shown in Table 1 [37,38]. The moisture contents of the biochar and pig manure compost were 5.5% and 27.2%, respectively. The biochar was generally alkaline with a pH of 9.7 and low in total nitrogen (TN), 2.0 g kg$^{-1}$.

**Table 1.** Chemical properties of biochar and pig manure compost used [1].

| Materials Used | pH | EC (dS m$^{-1}$) | TC (g kg$^{-1}$) | TOC (g kg$^{-1}$) | TIC (g kg$^{-1}$) | TN (g kg$^{-1}$) |
|---|---|---|---|---|---|---|
| Biochar | 9.67 ± 0.04 (1:10) | 1.4 ± 0.02 | 566 ± 5.2 | 533 ± 2.4 | 33.5 ± 0.8 | 2.0 ± 0.01 |
| Pig manure compost | 8.77 ± 0.02 (1:5) | 3.4 ± 0.03 | 289 ± 11.1 | 259 ± 20.7 | 30.2 ± 1.6 | 29.1 ± 0.3 |

[1] EC; Electric conductivity, TC; Total carbon, TOC; Total organic carbon, TIC; Total inorganic carbon, and TN; Total nitrogen. The values were average of triplicates samples with standard deviation.

### 2.2. Production of Supplemented Biochar Manure Pellet

The processing of SBMPFs is described in Figure 1. Prior to pelleting, biochar was processed in a series of sieves (0.5–5 mm) to ensure even particle distribution. In producing biochar pellets, 40% biochar was mixed with 60% pig manure compost as a binder. The SBMPF was completely mixed by using an agitator while spraying different nutrient solutions in the mixtures, and then feeding it into a commercial pellet mill (7.5 KW, 10HP, KumKang Engineering Pellet Mill Co., Daegu, South Korea). Different biochar pellets (Patent number: 10-1889400) treated with (1) urea solution heated at 60 °C (BMP-U60), (2) N, P, and K nutrient solutions at room temperature (BMP-NPK), (3) urea and K solutions at room temperature (BMP-UK), and (4) pig manure compost only (PMCP) pelletized. The size of BMPFs was approximately Ø 0.51 cm × 0.78 cm. The total carbon, TN (total nitrogen), TP (total phosphorus), and TK (total potassium) contents of BMPF embedded with different treatments are described in Table 2. Their total carbon and nitrogen contents varied from 225 g kg$^{-1}$ to 289 g kg$^{-1}$ and from 29.1 g kg$^{-1}$ to 102.0 g kg$^{-1}$, respectively. It was observed that the BMP-U60 had the highest nitrogen content of 102.0 g kg$^{-1}$ and BMP-UK had the lowest nitrogen content of 84.0 g kg$^{-1}$.

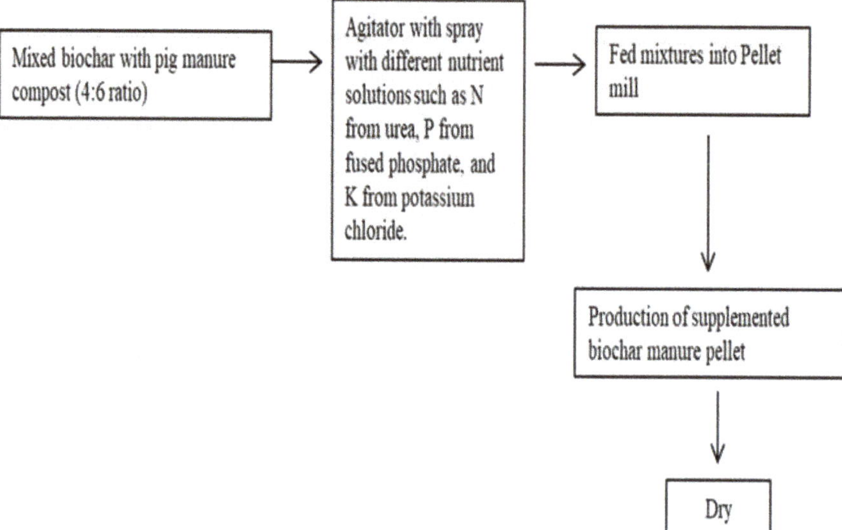


**Figure 1.** Diagram of processing the supplemented biochar manure pellets with different types of fertilizer.

**Table 2.** Total carbon, total nitrogen, total phosphorus and total potassium contents of supplemented biochar manure pellet fertilizers [1].

| Treatments * | TC (g kg$^{-1}$) | TN (g kg$^{-1}$) | TP(g kg$^{-1}$) | TK(g kg$^{-1}$) |
|---|---|---|---|---|
| PMCP | 289.0 ± 0.3 | 29.1 ± 0.01 | 79.4 ± 0.3 | 20.8 ± 0.2 |
| BMP-U60 | 226.3 ± 0.2 | 102.0 ± 0.25 | 29.5 ± 0.2 | 11.8 ± 0.3 |
| BMP-NPK | 227.8 ± 0.3 | 75.2 ± 0.03 | 32.8 ± 0.4 | 57.2 ± 0.3 |
| BMP-UK | 224.7 ± 0.5 | 84.0 ± 0.05 | 35.4 ± 0.3 | 13.5 ± 0.1 |

[1] TC; Total carbon, TN; Total nitrogen, TP; Total phosphorous, TK; Total potassium; * BMP-U60; BMP blended with urea solution heated at 60 °C, BMP-NPK; BMP blended with N, P and K nutrient solutions at room temperature and BMP-UK, BMP blended with N and P nutrient solutions at room temperature. The values displayed are averages of triplicate samples with standard deviation.

## 2.3. Field Experiment

The experimental field was cultivated with rice monoculture, and it has clay loamy soil. It is located at 35°49.510′ N of latitude and 127°2.536′ E of longitude in the National Institute of Agricultural Sciences (NIA), Rural Development Administration (RDA), Jeonju, Republic of Korea. The precipitation amount and average temperature were 718 mm and 22.3 °C during the rice cultivation season, respectively. Additionally, the solar radiation quantity and duration of sunshine are measured at 2753.2 MJ and 949.9 h during the cultivation period, respectively. The rice variety used in this experiment was Shindongjin, with a planting distance of 30 × 60 cm. The experimental design was a block design with five treatments consisting of (1) the control, (2) PMCP, (3) BMP-U60, (4) BMP-NPK, and (5) BMP-UK with three replications and 16 m$^2$ of the plot size. The amount of fertilizer and manure compost applied in the control and PMCP treatment were 90-45-57 kg ha$^{-1}$ (N-P-K) and 2600 kg ha$^{-1}$, respectively, which was based on National Institute of Agricultural Sciences (NIA) recommended rates for rice cultivation [40]. The SBMPFs were incorporated into the soil based on 90 N kg ha$^{-1}$ for whole basal application at 5 days prior to rice transplanting. Water logging time was 6 days prior to rice transplanting. The date of rice transplant was May 23, and drainage times were 14 days, 35 days, and 93 days after transplanting with one-week drainage. Rice was harvested 154 days after transplanting period. To evaluate the agricultural impact of different SBMPFs, major plant nutrients were analyzed from the surface water and soil in the paddy during rice cultivation. For rice growth responses, the plant height and number of tillers were measured about 100 days after rice transplanting, while the grain yield and dry weight of rice straw were weighed after harvest. For the effect of SBMPF applications in the paddy, the physicochemical properties of the soil used are presented in Table 3.

**Table 3.** Soil physicochemical properties of experimental field [1].

| Soil Type | pH | EC (dS m$^{-1}$) | NH$_4{}^+$–N (mg kg$^{-1}$) | NO$_3$–N (mg kg$^{-1}$) | P$_2$O$_5$ (mg kg$^{-1}$) | K$_2$O (mg kg$^{-1}$) | TC (g kg$^{-1}$) | TOC (g kg$^{-1}$) |
|---|---|---|---|---|---|---|---|---|
| Clay Loam | 7.0 ± 0.4 | 0.6 ± 0.03 | 10.6 ± 0.1 | ND | 97.8 ± 0.6 | 26.1 ± 0.1 | 20.7 ± 0.3 | 16.6 ± 0.2 |

[1] EC; electric conductivity, TC; Total carbon, TOC; Total organic carbon and ND; Non detected with 1 mg kg$^{-1}$ of detection limit. The values displayed are averages of triplicate samples with standard deviation.

### 2.4. Chemical Analysis of Paddy Soil and Water

After rice transplantation in the paddy, surface soil and water samples were collected every 20 days. The collected water samples were filtered through Whatman 2. The surface water was analyzed for NH$_4{}^+$–N, NO$_3{}^-$–N, K$^+$, and SiO$_2$ content using a UV spectrophotometer (C-Mac, Dae-Jeon, Korea) throughout the cropping season. The wet soil samples were extracted by using a 2M KCl solution (1:5, soil: extractant ratio). Those samples were analyzed directly for NH$_4{}^+$–N and NO$_3{}^-$–N by using the Bran-Lubbe Segmented Flow Auto Analyzer (Seal Analytical Ltd., Wisconsin, USA), and then the NH$_4{}^+$—N and NO$_3$—N concentrations were calculated by compensation for moisture contents of wet soil. The extractant using the Mehlich III method [41] from dried soil samples that passed through 2 mm sieves were stored in a refrigerator at 4 °C until PO$_4{}^-$, K$^+$ and SiO$_2$ were analyzed using a UV spectrophotometer (C-Mac, Dae-Jeon, Korea). Total carbon (TC) in soils was analyzed with total organic carbon (TOC) analyzer (Elementa vario TOC cube, Hanau, Germany). The combustion temperature was 950 °C and tungsten trioxide (WO$_3$) was used as the catalyst. With 350mg of soil samples, total nitrogen (TN) contents were determined by dry combustion with 250mg of L-Glutamic acid, standard compound, by using vario Max CN (Elementar, Hanau, Germany).

### 2.5. Data Processing and Carbon Balance Calculations

The soil carbon sequestration via BMPFs application was calculated from the difference of the residual amount of soil carbon between the control and different treatments after rice harvest by using the following equation [38]:

$$SS_{TC} = \left\{ \sum_{i=0}^{n} T_{TC} \, (Li - Ii) - NT_{TC} \, (Li - Ii) \right\} \times SW \tag{1}$$

where $SS_{TC}$ (kg ha$^{-1}$) is the potential sequestration amount of soil carbon, T (kg ha$^{-1}$) is the treatment of SBMPFs, NT (kg ha$^{-1}$) is the control, $_{TC}$ is total carbon content (g kg$^{-1}$), i is the sampling date, Li and Ii are carbon contents of the last and initial samplings which analyzed the soil carbon content (g kg$^{-1}$), and SW is the soil weight (bulk density, 1.3; 10cm of plowing soil depth, kg ha$^{-1}$).

The mitigation of CO$_2$ emission for SBMPFs application was also estimated using equation [38]:

$$CO_2 = SS_{TC} \times CF_{SC} \tag{2}$$

where $SS_{TC}$ is the amount of soil carbon sequestration (tonnes ha$^{-1}$) and $CF_{SC}$ is the conversion factor of CO$_2$ emission from soil carbon (1 kg C = 3.664 kg CO$_2$-equiv.).

Profit analysis for the mitigation of CO$_2$ emission was also calculated by using the equation [38]:

$$P = AM \times MP \tag{3}$$

where P is the profit of carbon dioxide trading ($ ha$^{-1}$), AM is the amount of mitigation of $CO_2$ emission (tonnes ha$^{-1}$), and MP is the market prices of $CO_2$ offsets ($ per tonnes $CO_2$). Also, the trading prices of $CO_2$ offsets in the European Climate Exchange (ECX) varied between $4.1 and $7.9 per tonnes $CO_2$ in 2016 [42] while the Korean Climate Exchange (KCX) ranged from $7.9 to $19.3 per 1 Korean Allowance Unit (KAU) [43].

*2.6. Statistical Analysis*

Statistical analysis was conducted using SAS version 9.2 Software (SAS, Inc., Cary, NC, USA), with an ANOVA with Duncan multiple range tests for the comparison of treatments with carbon contents at 1st day of rice transplanting and day after harvesting, carbon sequestration, and growth components during rice cultivation. Standard deviation was used for comparisons of paddy water and soil chemical properties.

## 3. Results and Discussions

*3.1. Effects of Essential Nutrients in the Paddy Water and Soil*

### 3.1.1. Paddy Water Quality

The $NH_4^+$–N and $NO_3^-$–N concentrations in the surface paddy water are presented in Figure 2. At the first day of rice transplanting, the $NH_4^+$–N concentration of surface paddy water in the MBP-NPK was significantly higher than the other treatments, but its control showed nearly the same values than the other treatments. However, the $NO_3^-$–N concentrations in the control and PMCP were only significantly higher than those in the SBMPF treatments. It was observed that $NH_4^+$–N concentrations in the treatments were higher on the first day of rice transplants, but similar to the rest of the days. The loss of nitrogen under the application of SBMPF was almost complete within 21 days after rice transplantation. This might be due to the adsorption of $NH_4^+$–N by the applied biochar in the soil. Regardless of the treatments at 112 days of rice transplanting, the $NO_3^-$–N concentrations were higher compared with other sampling days (93 days) due to the start of drainage of the surface water in the rice paddy. The study showed that the application of SBMPs can be a solution to mitigate the loss of nitrogen and phosphorus [44].

The $PO_4^-$–P, $K^+$, and $SiO_2$ concentrations in the surface paddy water under application of BMPFs are described in Figure 3. The measured $PO_4^-$–P concentration in the control and PMCP treatment was 2.8–5.3 times higher than the value in BMP-U60, BMP-UK, and BMP-NPK, respectively, until 21 days after rice transplantation. The $PO_4^-$–P concentrations were not significantly different ($p > 0.05$) from 41 days to 93 days after rice transplanting among the treatments. The greatest differences in K+ concentrations can be seen at 41 days after transplant. The higher values in the control and PMCP were 28.5 mg L$^{-1}$, and the lowest in the BMP-U60 was 9.6 mg L$^{-1}$, but not significantly different ($p > 0.05$) with that of BMP-UK.

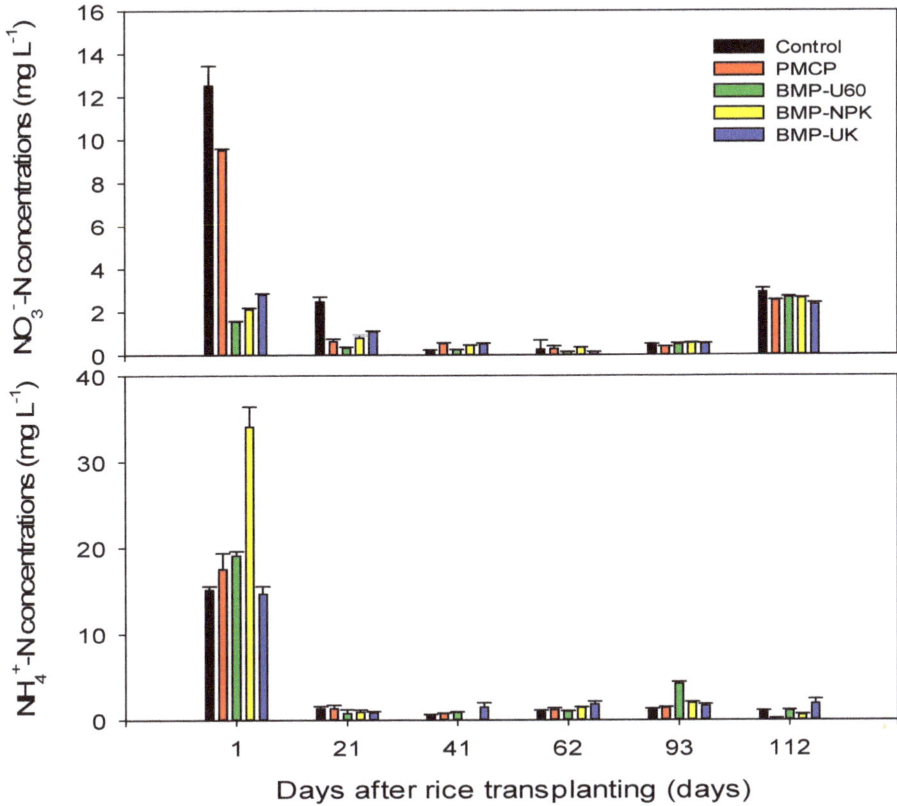

**Figure 2.** Effects of different treatments on $NH_4^+$–N and $NO_3^-$–N contents in rice surface paddy water during rice cultivation. The values displayed are averages of triplicate samples with standard deviation.

Silicon (Si) in soil exists in an unavailable form, but the Si in crop residues is a useful structure ($H_4SiO_4$) compared with Si fertilizer for crop uptake [45]. This recycled Si is leached into soil after the decomposition of crop residues. It is observed that $SiO_2$ concentration ranged from 10 mg $L^{-1}$ to 35 mg $L^{-1}$ during the cultivation period, and the highest $SiO_2$ concentration was 34.4 mg $L^{-1}$ in the BMP-UK at after 41 and 112 days of rice transplanting. However, $SiO_2$ concentrations in the paddy water under the application of SBMPFs were higher than those of the control and PMCP at 112 days after transplant. The most commonly used silicon fertilizer is wollastonite for soil application because of its high solubility for plant uptake (2.3–3.6%) [46]. Recently, much attention has been paid to biochar as an alternative soil ameliorant because it could slowly release 43 mg $kg^{-1}$ for the available plant uptake of silica [47]. The 1% KOH solution treated biochar application to soil significantly increased available form of silicon in the plant [48]. In this study, the $SiO_2$ concentration was significantly increased at the harvesting time under the application of SBMPFs. Thus, the incorporation of SBMPFs had the potential ability to recycle silica. Overall, the $PO_4^-$P, $K^+$, and $SiO_2$ concentrations were significantly higher than the other sampling days (93 days) due to the start of drainage of the surface water in the paddy field.

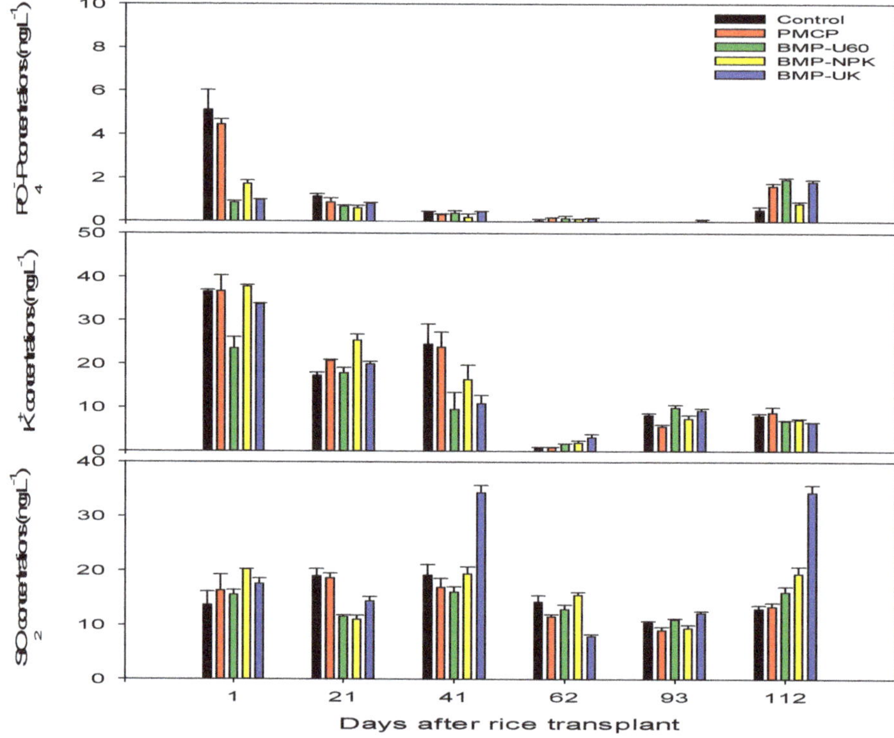

**Figure 3.** Effects of different treatments on $PO_4^--P$, $K^+$ and $SiO_2$ concentrations in surface paddy water during rice cultivation. The values displayed are averages of triplicate samples with standard deviation.

### 3.1.2. Nutrients in Paddy Soil

Urea application is usually the main source of ammonium ions because urea can be hydrolyzed into $NH_4^+$ and $OH^-$ by the ammonification reaction within short periods after application in the paddy soil. The major nutrient concentrations in the soil are described in Figure 4. $NH_4^+-N$ concentration in the BMP-NPK was highest among the treatments at 41 days after rice transplanting. Total nitrogen losses were reduced with the incorporation of rice straw in the rice paddy soil due to increasing immobilization [49] and denitrification [50]. $P_2O_5$ concentrations except the PMCP were not significantly different during 21 days after rice transplanting among treatments. The $K_2O$ concentrations in the soil treated with BMPFs continuously decreased during rice cultivation due to the $K^+$ solubility, except for the BMP-U60 treatment. Biochar application increased the availability of $K^+$ and P because it was a net source of cations due to increased soil capacity to hold exchangeable cations [51,52]. The application of biochar produced from rice straw increased the available P and $K^+$ by 15.3% and 28.6% in the soil, respectively. However, biochar application did not significantly increase total nitrogen compared with the control in the rice paddy [53]. Overall, the release of major nutrients to soil under the application of SBMPFs was significantly lower compared with those from the control and PMCP.

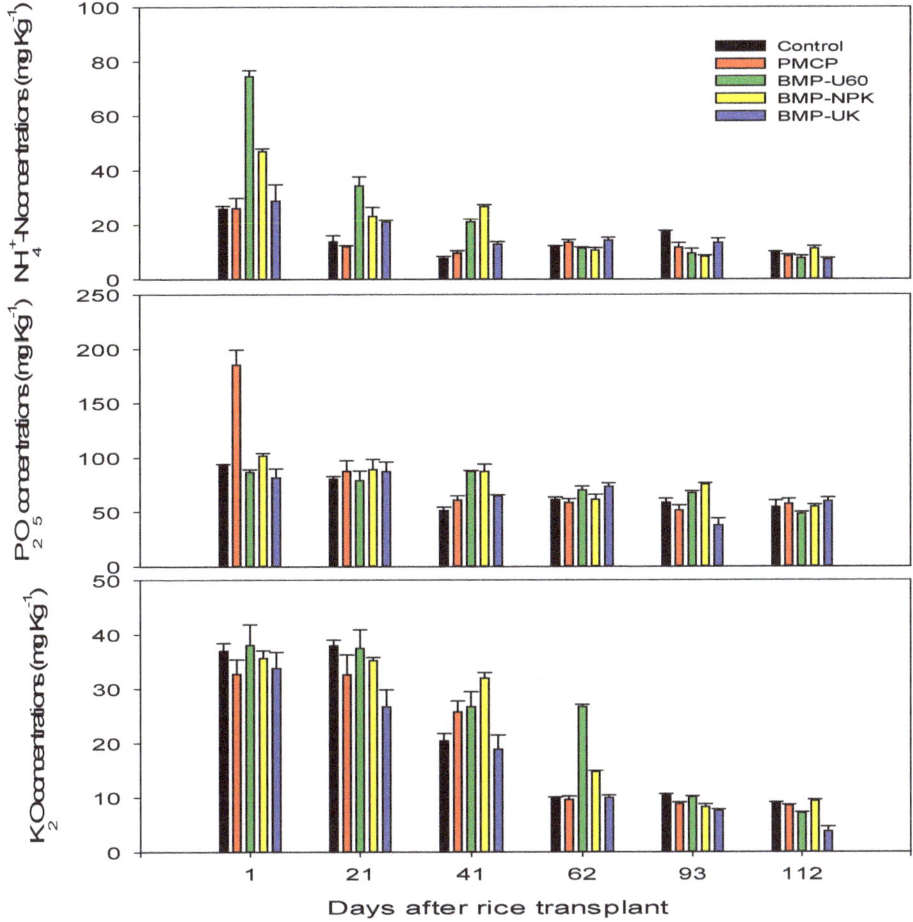

**Figure 4.** $NH_4^+$–N, $P_2O_5$ and $K_2O$ concentrations under different treatments in the paddy soil during rice cultivation. The values displayed are averages of triplicate samples with the standard deviation.

## 3.2. Carbon Sequestration and Profit Analysis

Soil carbon sequestration was only considered after soil analysis from rice paddy incorporated SBMPFs at day 1 of rice transplanting and the day after harvesting. Changes of total carbon contents in paddy soil under different treatments at the initial stage and after harvesting are described in Table 4. The carbon contents on first day of rice transplanting and the day after harvesting were significantly ($p < 0.001$) different in the treatments. There was minimal difference in total carbon content in the control between the first day of rice transplanting and after harvesting.

The application of biochar incorporated to the soil has been suggested as a promising method for carbon sequestration as well as another method for mitigating greenhouse gas, increasing crop yields and enhancing the sorption of pollutants [49,54]. Regarding carbon sequestration, it might be distinguished that short term released $CO_2$ refers to the retention time of sequestrated carbon in soil from organic matter decomposition, while long term, it is stored as biochar from thermal conversion materials [38].

**Table 4.** Carbon contents in the soils treated with different supplemented biochar manure pellet fertilizers on first day of rice transplant and day after harvest *.

| Treatments Control | First Day of Rice Transplant (g kg$^{-1}$) | Day After Harvest (g kg$^{-1}$) |
|---|---|---|
|  | 10.30 ± 0.02 a | 10.38 ± 0.02 c |
| PMCP | 9.45 ± 0.07 d | 10.49 ± 0.07 c |
| BMP-U60 | 9.87 ± 0.13 b | 10.83 ± 0.13 b |
| BMP-NPK | 9.90 ± 0.06 b | 11.80 ± 0.09 a |
| BMP-UK | 9.66 ± 0.05 c | 11.83 ± 0.03 a |
| F-value | 55.33 | 235.30 |
| Pr > F | <0.001 | <0.001 |

* Mean values followed by different letters, which indicate significant differences ($p < 0.05$) among treatments with One way ANOVA by the mean comparison for all pairs using Tukey-Kramer HSD analysis for total carbon contents on first day of rice transplant and the day after harvest.

For the application of different types of SBMPFs, carbon sequestration, mitigation of $CO_2$, and profit analysis were calculated by using Equations (1)–(3), respectively (Table 5). The analysis of carbon sequestration showed 2.67 tonnes C ha$^{-1}$ in the BMP-UK as the best treatment for carbon sequestration, and 1.14 tonnes C ha$^{-1}$ in the BMP-U60 as the worst. It appeared that their recovery rates varied from 25.4% to 48.5% of SBMPFs applied to the rice paddy. It was observed that the mitigation of $CO_2$ increased with the application of BMPFs, and the highest was 5.09 tonnes C ha$^{-1}$ in the BMP-UK. The profit under SBMPFs application was estimated to range from \$6.56 ha$^{-1}$ to \$68.80 ha$^{-1}$ during rice cultivation for KAU. The target of the Korean government is to reduce greenhouse gas emissions by 1.48 million tonnes $CO_2$-equiv. (5.2%) of the 28.49 million tonnes $CO_2$-equiv. total greenhouse emissions in the agricultural sector by 2020 [55]. Therefore, it is estimated that the 482,085 ha$^{-1}$ (29.3%) of 1,644,000 ha$^{-1}$ total area of rice cultivation with the BMP-NPK application in Korea [56] is required to accomplish this goal.

In order to establish carbon trading in the agriculture sector, policymakers should prepare a draft policy specifically for mitigating greenhouse gas emissions by providing support to farmers of about \$58 per hectare of cultivated rice paddy through the application of BMP-NPK. The application of BMPFs did not only increase carbon storage, but also enhanced rice yield and soil fertility [38].

**Table 5.** Evaluation of carbon sequestration and its profit analysis for application of supplemented biochar manure pellet fertilizers during rice cultivation.

| Treatments | Carbon Sequestration (Tonnes ha$^{-1}$) | Mitigation of $CO_2$ (Tonnes ha$^{-1}$) | Profit (\$ ha$^{-1}$) | Additional Profit for SBMPF Application (\$ ha$^{-1}$) |
|---|---|---|---|---|
| Control | 1.28 ± 0.11 b | 4.70 ± 0.12 b | 63.59 ± 2.50 b | - |
| PMCP | 1.24 ± 0.08 b | 4.54 ± 0.29 b | 61.47 ± 3.96 b | - |
| BMP-U60 | 1.41 ± 0.12 b | 5.18 ± 0.44 b | 70.06 ± 5.98 b | 6.56 |
| BMP-NPK | 2.45 ± 0.18 a | 8.98 ± 0.66 a | 121.46 ± 8.92 a | 57.87 |
| BMP-UK | 2.67 ± 0.12 a | 9.78 ± 0.44 a | 132.36 ± 5.95 a | 68.77 |
| F-value | 55.06 | 55.06 | 55.06 | - |
| Pr > F | <0.001 | <0.001 | <0.001 | - |

kg C = 3.664 kg $CO_2$-eqiv., 1 tonnes $CO_2$ = KAU = 23,000 (8.12) = \$13.53.

### 3.3. Rice Growth Responses to Supplemented Biochar Manure Pellet

Growth responses to the application of SBMPFs are shown in Table 6. The plant height in BMP-U60 was 15.2% higher than the control, and rice yield in the BMP-U60 was increased by 15.7% compared with the control, even when the application amount of pig manure compost applied was reduced to about 1000 kg ha$^{-1}$. This result might be due to the enhanced nutrient use efficiency under application of BMPFs functioning as a slow release fertilizer. Min et al. [4] reported that supplemented BMPFs application enhanced rice yield. Shin et al. [38] also reported similar results in their study. With the whole basal application of SBMPFs in the rice field prior to rice transplanting, it could prevent additional fertilizer application. Puga et al. [57] conducted similar research to evaluate the effects of biochar-based N fertilizers on nitrogen use efficiency (NUE) and maize yield. Their results showed that an average maize yield was increased 26% in the application of biochar-based N fertilizers (51% biochar with 10% N) compared with urea only treatment, and the NUE was 12% improved. Pokharel and Chang [58] also reported that manure pellet with wood chip biochar significantly increased plant grain yield by 36.3 and 16.1%, compared to the control, while woodchip with biochar applications significantly decreased plant grain yield.

**Table 6.** Characteristics of rice growth to supplemented biochar manure pellet fertilizer application.

| Treatments | Plant Height (cm) | Number of Tillers | Dry Weight of Rice Straw (Tonnes ha$^{-1}$) | Grain Yield (Tonnes ha$^{-1}$) |
|---|---|---|---|---|
| Control | 92.33 ± 0.58 b | 11.67 ± 1.53 b | 9.73 ± 0.51 a | 6.63 ± 0.14 b |
| PMCP | 100.00 ± 2.00 ab | 12.33 ± 2.52 ab | 9.55 ± 0.11 a | 6.68 ± 0.49 ab |
| BMP-U60 | 106.33 ± 8.15 a | 16.00 ± 2.65 ab | 6.85 ± 0.43 b | 7.67 ± 0.36 a |
| BMP-NPK | 103.67 ± 5.51 ab | 13.00 ± 3.46 ab | 5.96 ± 0.51 c | 7.13 ± 0.33 a |
| BMP-UK | 104.67 ± 5.03 ab | 17.67 ± 3.51 a | 5.32 ± 0.53 c | 6.52 ± 0.65 b |
| F-value | 3.69 | 2.49 | 63.02 | 3.69 |
| Pr > F | 0.043 | 0.110 | <0.001 | 0.043 |

## 4. Conclusions

Different supplemented biochar manure pellet fertilizers were tested to assess their agro-environmental impacts on paddy water and soil systems during rice cultivation. With regard to the water quality of paddy, the $NO_3^--N$ and $PO_4^--P$ in control and PMCP were relatively higher than those of the SBMPFs applied plots. Non-point pollutants in runoff water to small stream near the rice cultivation area were reduced with application of SBMPFs. Considering the soil chemical properties, $NH_4^+-N$ concentration in control was lower compared with the SBMPFs treatment at 41 days after rice transplant. However, the available $P_2O_5$ concentrations were almost stage-state among all the treatments from 21 days after rice plant until the harvest period, except for the first day of rice transplant in the PMCP. It is possible that the SBMPFs can be applied with whole basal application without additional application of chemical fertilizers. Also, the highest carbon sequestration was 2.67 tonnes C ha$^{-1}$ in BMP-UK treatment, and the lowest was 1.14 tonnes C ha$^{-1}$ in the BMP-U60 treatment. The grain yields from the SBMPF applied plots, except for BMP-UK, were significantly higher than the yield from the control even though amounts of pig manure compost applied were decreased from 1881.8 kg ha$^{-1}$ to 2070.8 kg. Therefore, the application of SBMPFs can contribute to reducing the agro-environmental impacts of runoff as well as enhance carbon sequestration and rice yield in agro-ecosystems.

**Author Contributions:** Project leader and original draft writing, J.S.; Statistics and visualization, S.P.; review and editing, C.J. All authors have read and agreed to the published version of the manuscript.

**Funding:** This research was funded beyond Research Program of Agricultural Science & Technology Development (Project No. PJ013814012020) in Korea.

**Acknowledgments:** We are thankful to the National Institute of Agricultural Sciences, Rural Development Administration in Korea.

**Conflicts of Interest:** The author certifies that there are no affiliation with or involvement in any organization or entity with any financial interest (such as honoraria; educational grants; participation in speakers' bureaus; membership, employment, consultancies, stock ownership, or other equity interest; and expert testimony or patent-licensing arrangements), or non-financial interest (such as personal or professional relationships, affiliations, knowledge or beliefs) in the subject matter or materials discussed in this manuscript.

## References


1. Edgerton, M.D. Increasing crop productivity to meet global needs for feed, food, and fuel. *Plant Physiol.* **2009**, *149*, 7–13. [CrossRef] [PubMed]
2. O'Connor, D.; Peng, T.; Zhang, J.; Tsang, D.C.W.; Alessi, D.S.; Shen, Z.; Bolan, N.S.; Hou, D. Biochar application for the remediation of heavy metal polluted land: A review of in situ field trials. *Sci. Total Environ.* **2018**, *619*, 815–826. [CrossRef] [PubMed]
3. Zhao, B.; O'Connor, D.; Zhang, J.; Peng, T.; Shen, Z.; Tsang, D.C.W.; Hou, D. Effect of pyrolysis temperature, heating rate, and residence time on rapeseed stem derived biochar. *J. Clean. Prod.* **2018**, *174*, 977–987. [CrossRef]
4. Min, H.; Liu, Y.; Qin, H.; Jiang, L.; Zou, Y. Quantifying the effect of biochar amendment on soil quality and crop productivity in Chinese rice paddies. *Field Crops Res.* **2013**, *154*, 172–177.
5. Li, Z.; Gu, C.; Zhang, R.; Ibrahim, M.; Zhang, G.; Wang, L.; Zhang, R.; Chen, F.; Liu, Y. The benefic effect induced by biochar on soil erosion and nutrient loss of slopping land under natural rainfall conditions in central China. *Agric. Water Manag.* **2017**, *185*, 145–150. [CrossRef]
6. Liu, X.; Zhou, J.; Chi, Z.; Zheng, J.; Li, L.; Zhang, X.; Zheng, J.; Cheng, K.; Bian, R.; Pan, G. Biochar provided limited benefits for rice yield and greenhouse gas mitigation six year following an amendment in a rice paddy. *Catena* **2019**, *179*, 20–28. [CrossRef]
7. MIFAFF. Annual statistics in food, agriculture, fisheries and forestry in 2009. Korean Ministry for Food, Agriculture, Fisheries and Forestry. *Environ. Sci. Pollut. Res.* **2010**, *16*, 1–9.
8. Lehmann, J.; Rillig, M.C.; Thies, J.; Masiello, C.A.; Hockaday, W.C.; Crowley, D. Biochar effects on soil biota —A review. *Soil Biol. Biochem.* **2011**, *43*, 1812–1836. [CrossRef]
9. Cantrell, K.B.; Hunt, P.G.; Uchimiya, M.; Novak, J.M.; Ro, K.S. Impact of pyrolysis temperature and manure source on physicochemical characteristics of biochar. *Bioresour. Technol.* **2012**, *107*, 419–428. [CrossRef] [PubMed]
10. Brewer, C.; Unger, R.; Schmidt-Rohr, K.; Brown, R. Criteria to select biochars for field studies based on biochar chemical properties. *Bioenergy Res.* **2011**, *4*, 312–323. [CrossRef]
11. Xie, T.; Reddy, K.R.; Wang, C.; Yargicoglu, E.; Spokas, K. Characteristics and applications of biochar for environmental remediation: A review. *Crit. Rev. Env. Sci. Tech* **2015**, *45*, 939–969. [CrossRef]
12. Kim, P.; Johnson, A.M.; Essington, M.E.; Radosevich, M.; Kwon, W.T.; Lee, S.H.; Rials, T.G.; Labbe, N. Effect of pH on surface characteristics of switchgrass-derived biochars produced by fast pyrolysis. *Chemosphere* **2013**, *90*, 2623–2630. [CrossRef] [PubMed]
13. Blackwell, P.; Joseph, S.; Munroe, P.; Anawar, H.M.; Storer, P.; Gilkes, R.J.; Solaiman, Z.M. Influences of biochar and biochar-mineral complex on mycorrhizal colonization and nutrition of wheat and sorghum. *Pedosphere* **2015**, *25*, 686–695. [CrossRef]
14. Chan, K.; van Zwieten, L.; Meszaros, I.; Downie, A.; Joseph, S. Agronomic values of green waste biochar as a soil amendment. *Aust. J. Soil Res.* **2007**, *45*, 629–634. [CrossRef]
15. Deenik, J.L.; McClellan, T.; Uehara, G.; Antal, M.J.; Campbell, S. Charcoal volatile matter content influences plant growth and soil nitrogen transformations. *Soil Sci. Soc. Am. J.* **2010**, *74*, 1259–1270. [CrossRef]
16. Solaiman, Z.M.; Murphy, D.V.; Abbott, L.K. Biochars influence seed germination and early growth of seedlings. *Plant Soil* **2012**, *353*, 273–287. [CrossRef]
17. Husk, B.; Major, J. *Commercial Scale Agricultural Biochar Field Trial in Quebec, Canada, over Two Years: Effects of Biochar on Soil Fertility, Biology, Crop Productivity and Quality*; Blue Leaf: Quebec, QC, Canada, 2008.
18. Major, J.; Lehmann, J.; Rondon, M.; Goodale, C. Fate of soil-applied black carbon: Downward migration, leaching and soil respiration. *Glob. Chang. Biol.* **2010**, *16*, 1366–1379. [CrossRef]
19. Reza, M.T.; Lynam, L.G.; Vasquez, V.R.; Coronella, C.J. Pelletization of biochar from hydrothermally carbonized wood. *Environ. Prog. Sustain. Energy* **2012**, *31*, 225–234. [CrossRef]
20. Khalil, M.; Gutser, R.; Schmidhalter, U. Effects of urease and nitrification inhibitors added to urea on nitrous oxide emissions from a loess soil. *J. Plant Nutr. Soil Sci.* **2009**, *172*, 651–660. [CrossRef]
21. Harmel, R.D.; Torbert, H.A.; Haggard, B.E.; Haney, R.; Dozier, M. Water quality impacts of converting to a poultry litter fertilization strategy. *J. Environ. Qual.* **2004**, *33*, 2229–2242. [CrossRef]
22. Wang, Y.; Lin, Y.; Chiu, P.C.; Imhoff, P.T.; Guo, M. Phosphorus release behaviors of poultry litter biochar as a soil amendment. *Sci. Total Environ.* **2015**, *512*, 454–463. [CrossRef] [PubMed]

23. EPA. A Long-Term Vision for Assessment, Restoration, and Protection under the Clean Water Act Section 2013, 303(d) Program. Available online: https://19january2017snapshot.epa.gov/new-version-cwa-303d-program-updated-framework-implementing-cwa-303d-program-responsibilities_html (accessed on 14 April 2020).

24. Cantrell, K.B.; Martin, J.H., II. Poultry litter and switchgrass blending and pelleting characteristics for biochar production. In Proceedings of the ASABE Annual International Meeting, Dallas, TX, USA, 29 July –1 August 2012.

25. Hua, L.; Wu, W.X.; Liu, Y.X.; McBride, M.; Chen, Y.X. Reduction of nitrogen loss and Cu and Zn mobility during sludge composting with bamboo charcoal amendment. *Environ. Sci. Pollut. Res.* **2009**, *16*, 1–9. [CrossRef] [PubMed]

26. Ro, K.S.; Cantrell, K.B.; Hunt, P.G. High-temperature pyrolysis of blended animal manures for producing renewable energy and value-added biochar. *Ind. Eng. Chem. Res.* **2010**, *49*, 10125–10131. [CrossRef]

27. Faloye, O.T.; Alatise, M.O.; Ajayi, A.E.; Ewulo, B.S. Synergistic effects of biochar and inorganic fertilizer on maize (zea mays) yield in an alfisol under drip irrigation. *Soil Tillage Res.* **2017**, *174*, 214–220. [CrossRef]

28. Fernandez-Escobar, R.; Benlloch, M.; Herrera, E.; Garcia-Novelo, J.M. Effect of traditional and slow-release N fertilizers on growth of olive nursery plants and N losses by leaching. *Sci. Hortic.* **2004**, *101*, 39–49. [CrossRef]

29. Mortain, L.; Dez, I.; Madec, P.J. Development of new composites materials, carriers of active agents from biodegradable polymers and wood. *Comptes Rendus Chim.* **2004**, *7*, 635–640. [CrossRef]

30. Kim, J.; Yoo, G.; Kim, D.; Ding, W.; Kang, H. Combined application of biochar and slow-release fertilizer reduces methane emission but enhances rice yield by different mechanisms. *Appl. Soil Ecol.* **2017**, *117*, 57–62. [CrossRef]

31. Zhao, L.; Cao, X.; Zheng, W.; Kan, Y. Phosphorus-assisted biomass thermal conversion: Reducing carbon loss and improving biochar stability. *PLoS ONE* **2014**, *9*, e115373. [CrossRef]

32. Brassard, P.; Godbout, S.; Raghavan, V. Soil biochar amendment as a climate change mitigation tool: Key parameters and mechanisms involved. *J. Environ. Manage.* **2016**, *181*, 484–497. [CrossRef]

33. Harvey, O.R.; Kuo, L.J.; Zimmerman, A.R.; Louchouarn, P.; Amonette, J.E.; Herbert, B.E. An index-based approach to assessing recalcitrance and soil carbon sequestration potential of engineered black carbons (biochars). *Environ. Sci. Technol.* **2012**, *46*, 1415–1421. [CrossRef]

34. Cross, A.; Sohi, S.P. A method for screening the relative long term stability of biochar. *Gcb Bioenergy* **2013**, *5*, 215–220. [CrossRef]

35. Lehmann, J.; Joseph, S. *Biochar for Environmental Management: Science, Technology and Implementation*; Routledge: New York, NY, USA, 2015.

36. Shin, J.; Hong, S.; Lee, S.; Hong, S.; Lee, J. Estimation of soil carbon sequestration and profit analysis on mitigation of $CO_2$-eq. emission in cropland incorporated with compost and biochar. *Appl. Biol. Chem.* **2017**, *60*, 467–472. [CrossRef]

37. Shin, J.; Choi, E.; Jang, E.S.; Hong, S.G.; Lee, S.; Ravindran, B. Adsorption characteristics of ammonium nitrogen and plant responses to biochar pellet. *Sustainability* **2018**, *10*, 1331. [CrossRef]

38. Shin, J.; Jang, E.; Park, S.; Ravindran, B.; Chang, S. Agro-environmental impacts, carbon sequestration and profit analysis of blended biochar pellet application in the paddy soil-water system. *J. Environ. Manag.* **2019**, *244*, 92–98. [CrossRef] [PubMed]

39. Thammasom, N.; Vityakon, P.; Lawongsa, P.; Saenjan, P. Biochar and rice straw have different effects on soil productivity, greenhouse gas emission and carbon sequestration in Northeast Thailand paddy soil. *Agric. Nat. Resour.* **2016**, *50*, 192–198. [CrossRef]

40. NAAS. *Recommended Application Amounts of Fertilizers for Crop Cultivation (eds)*; National Academy of Agricultural Sciences, Rural Development Administration: New Delhi, India, 2010; p. 16.

41. Mehlich, A. Mehlich III soil test extractant: A modification of Mehlich II extractant. *Commun. Soil Sci. Plant Anal.* **1984**, *15*, 1409–1416. [CrossRef]

42. EEX. EU Emission Allowances—Primary Market Auction. European Energy Exchange. 2017. Available online: https://www.eex.com/en/marketdata/environmental-markets/auction-market/ (accessed on 7 March 2017).

43. KRX. Korea Allowance Unit. Korean Exchange. Available online: http://marketdata.krx.co.kr/mback; http://open.krx.co.kr/contents/OPN/01/01050401/OPN01050401.jsp#document=070301 (accessed on 7 March 2017).

44. Kaneki, R.; Iwama, K.; Minagawa, A.; Sudo, M.; Odani, H.; Kobayashi, A.; Tanaka, A.; Ikeda, K.; Muranaga, M. Effect of mass balance in paddy fields and rice plant yields by reduced fertilizer use and non-puddling cultivation. *J. Jpn. Soc. Hydrol. Water Resour.* **2013**, *26*, 201–211. [CrossRef]

45. Epstein, F. Silicon: Its manifold roles in plants. *Ann. Appl. Biol.* **2009**, *155*, 155–160. [CrossRef]
46. Wang, J.; Wang, D.; Zhang, G.; Wang, Y.; Wang, C.; Teng, Y.; Christie, P. Nitrogen and phosphorous leaching losses from intensively managed paddy fields with straw retention. *Agri. Water Manag.* **2014**, *141*, 66–73. [CrossRef]
47. Hasegawa, K.; Kobayashi, M.; Nakada, H. Influence of applying organic matter in a paddy field on the water quality (2). Influence of applying rice-straw on the leaching of nitrate nitrogen from draining soil and on the denitrification of nitrate nitrogen in flooded soil. *Bull. Shiga. Agric. Exp. Stn.* **1981**, *23*, 30–37.
48. Xu, G.; Wei, L.L.; Sun, J.N.; Shao, H.B.; Chang, S.X. What is more important for enhancing nutrient bioability with biochar application into a sandy soil: Direct or indirect mechanism? *Ecol. Eng.* **2013**, *52*, 119–124. [CrossRef]
49. Major, J.; Steiner, C.; Downie, A.; Lehmann, J. Biochar effects on nutrient leaching. In *Biochar for Environmental Management: Science and Technology*; Lehmann, J., Joseph, S., Eds.; Earchscan Publ.: London, UK, 2009; pp. 271–287.
50. Liu, Y.; Lu, H.; Yang, S.; Wang, Y. Impacts of biochar addition on rice yield and soil properties in a cold waterlogged paddy for two crop seasons. *Field Crops Res.* **2016**, *191*, 161–167. [CrossRef]
51. Sebastian, D.; Rodrigues, H.; Kinsey, C.; KorndÖrfer, G.; Pereira, H.; Buck, G.; Datnoff, L.; Miranda, S.; Provance-Bowely, M. A 5-day method for determination of soluble silicon concentrations in non-liquid fertilizer materials using a sodium carbonate-ammonium nitrate extractant followed by visible spectroscopy with heteropoly blue analysis; single-laboratory validation. *J. AOAC Int.* **2013**, *96*, 251–259. [CrossRef] [PubMed]
52. Xiao, X.; Chen, B.; Zhu, L. Transformation, morphology, and dissolution of silicon and carbon in rice straw-derived biochars under different pyroltytic temperatures. *Environ. Sci. Technol.* **2014**, *28*, 3411–3419. [CrossRef] [PubMed]
53. Wang, M.; Wang, J.; Wang, X. Effect of KOH-enhanced biochar on increasing soil plant-available silicon. *Geoderma* **2018**, *321*, 22–31. [CrossRef]
54. Kan, T.; Strezov, V.; Evans, T.J. Lignocellulosic biomass pyrolysis: A review of product properties and effects of pyrolysis parameters. *Renew. Sustain. Energy Rev.* **2016**, *57*, 1126–1140. [CrossRef]
55. MAFFa. *Report of Road Map to Accomplish the Reduction Goal of Greenhouse Gas Emissions in Korea*; Korean Ministry for Food, Agriculture, Fisheries and Forestry: Gwacheon, Korea, 2014; pp. 57–60.
56. MAFFb. *Annual Statistics in Food, Agriculture, Fisheries and Forestry in 2015*; Korean Ministry for Food, Agriculture, Fisheries and Forestry: Sejong, Korea, 2015.
57. Puga, A.P.; Grutzmacher, P.; Cerri CE, P.; Ribeirinho, V.S.; de Andrade, C.A. Biochar-based nitrogen fertilizers: Greenhouse gas emissions, use efficiency, and maize yield in tropical soils. *Sci. Total Environ.* **2020**, *704*, 135375. [CrossRef]
58. Pokharel, P.; Chang, S.X. Manure pellet, woodchip and their biochars differently affect wheat yield and carbon dioxide emission from bulk and rhizosphere soils. *Sci. Total Environ.* **2019**, *659*, 463–472. [CrossRef]

MDPI

*Article*

# Natural Grasslands as Lignocellulosic Biofuel Resources: Factors Affecting Fermentable Sugar Production


Linda Mezule [1,*], Baiba Strazdina [2], Brigita Dalecka [1], Eriks Skripsts [3] and Talis Juhna [1]

[1] Water Research and Environmental Biotechnology Laboratory, Riga Technical University, P. Valdena 1-303, LV-1048 Riga, Latvia; brigita.dalecka_1@rtu.lv (B.D.); talis.juhna@rtu.lv (T.J.)
[2] Latvian Fund for Nature, Vilandes 3-7, LV-1010 Riga, Latvia; baiba.strazdina@ldf.lv
[3] Bio RE LTD, Vadzu 34, LV-1024 Riga, Latvia; eriks.skripsts@biore.lv
* Correspondence: linda.mezule@rtu.lv



**Abstract:** Semi-natural grassland habitats are most often limited to animal grazing and low intensity farming. Their potential in bioenergy production is complicated due to the heterogeneity, variation, accessibility, and need for complex pre-treatment/hydrolysis techniques to convert into valuable products. In this research, fermentable sugar production efficiency from various habitats at various vegetation periods was evaluated. The highest fermentable sugar yields (above 0.2 g/g volatile solids) over a period of 3 years were observed from habitats "xeric and calcareous grasslands" (Natura 2000 code: 6120) and "semi-natural dry grasslands and scrubland facies on calcareous substrates" (Natura 2000 code: 6210). Both had a higher proportion of dicotyledonous plants. At the same time, the highest productivity (above 0.7 t sugar/ha) was observed from lowland hay meadows in the initial stage of the vegetation. Thus, despite variable yield-affecting factors, grasslands can be a potential resource for energy production.

**Keywords:** fermentable sugar; enzymatic hydrolysis; lignocellulosic biomass

**Citation:** Mezule, L.; Strazdina, B.; Dalecka, B.; Skripsts, E.; Juhna, T. Natural Grasslands as Lignocellulosic Biofuel Resources: Factors Affecting Fermentable Sugar Production. *Energies* **2021**, *14*, 1312. https://doi.org/10.3390/en14051312

Academic Editor: Alberto Coz

Received: 4 January 2021
Accepted: 23 February 2021
Published: 28 February 2021

**Publisher's Note:** MDPI stays neutral with regard to jurisdictional claims in published maps and institutional affiliations.

## 1. Introduction

Worldwide attention towards application of waste materials for energy and high-value chemical production has become a standard. Extensive use of agricultural and wood processing waste in lignocellulosic biofuel production increases the overall turnover of this industry annually. Furthermore, the use of lignocellulosic biomass for biofuel production is now facilitated by the European Union (EU) Renewable Energy Directive 2018/2001 [1]—the resource is included as alternative raw material under Annex IX. Regrettably, biomass recalcitrance towards saccharification is often the major limitation in the conversion of the resource to valuable end-products. Effective and economically feasible extraction of fermentable sugars is closely linked to the selection of an appropriate pre-treatment/hydrolysis technique and to the type of biomass used. A tremendous amount of studies have been performed to evaluate the potential of certain biomass resources, e.g., wheat or barley straw, corn stover, with various technologies and their combinations [2,3], resulting in an extensive amount of data and laboratory scale research. Furthermore, it has been demonstrated that the combination of climate, soil fertility, and grassland biomass type can influence the overall bioenergy potential, i.e., hydrolysis efficiency and fermentable sugar yields [2,4].

Currently in the EU, more than 61 million hectares are occupied by permanent grasslands [5] where temperate semi-natural grasslands with a long extensive management history represent the richest species ecosystems on earth. At a small spatial scale, their vascular plant diversity exceeds tropical rainforests, which are normally considered as global maxima [6]. Ref. [7] described the trend of grassland management abandonment due to economic reasons in Europe, leaving huge amounts of this resource unused. The abandoned areas are predominantly semi-natural and nature conservation grasslands, bearing a large variety of plant and animal species. Most of these grasslands are characterized by low productivity, but the optimal management regime includes low-intensity agricultural practices. In many cases, this means controlled grazing or late seasonal harvest that leads to the creation of patchiness, selection of particular species, or high amounts of lignin and cellulose in the biomass, respectively. Thus, forage quality is reduced [8,9]. Therefore, it is necessary to find alternative management regimes to maintain the biodiversity in European manmade landscapes [10] and at the same time to facilitate sustainable use of this resource. Unfortunately, semi-natural grasslands cannot be evaluated on a species level due to the high diversity and variability of the vegetation. Species composition and, especially, the coverage and distribution of particular species can vary even within one vegetation class or small grassland plot. Furthermore, it is influenced by environmental conditions [11–13], management [14,15], surrounding areas [16], land use history [17], and other factors. Thus, it is crucial to investigate and perform proper evaluation of the local grassland variations, their productivity and variability to estimate the costs and possible yields of biomass that can be further converted into high value chemicals, including biofuels [18].

Grass co-digestion with other waste streams to produce biogas has been shown to be efficient [19]. It is estimated that 8–17% of the current grassland biomass could provide up to 1% of EU transport fuel [20]. However, the high effect of area-specific biomass diversity, cutting time, accessibility, and need for pre-treatment have limited the potential use of grass in biogas production at an industrial level [21,22]. As an alternative to methane production via complex anaerobic digestion process, the use of lignocellulosic grassland biomass has been demonstrated for fermentable sugar production [23], which is an intermediate stage to produce various liquid biofuels, e.g., bioethanol or biobutanol, high value chemicals, used as an additional feedstock in biogas stations or regarded as a first step towards biorefinery [20]. The aim of this study was to evaluate fermentable sugar yields and overall productivity potential from various grassland habitats that are common in a temperate climate and classified under EU habitat codes. To aid towards biorefinery, non-commercial enzymes extracted from white rot fungi were used in the hydrolysis. The assessment involved not only the evaluation of habitat type but also seasonality, cutting time, species diversity, and solid content in the biomass. In-house made enzymes were preferred to commercial products due to their potential onsite production capacity and, thus, minimization of manufacturing costs. To the best of the authors' knowledge, this is the first study where the Natura 2000 grassland habitat classification [24] is linked with fermentable sugar productivity in the Baltic region, thus offering new grassland management practices by facilitating the of use of these resources for high value chemical production.

## 2. Materials and Methods

### 2.1. Biomass Sampling

In total, 162 grass biomass samples were collected from 67 randomly selected semi-natural grassland plots in Sigulda and Ludza municipalities (Latvia) over a 3 year period (Supplementary Materials Annex 1), corresponding to 6 habitat types of Community importance (the most common habitat types within these municipalities), and classified under the EU (Table 1).

**Table 1.** Description of analyzed semi-natural grassland habitats.

| European Union (EU) Habitat Type [24] | National Variants of EU Habitat Type [25] | PAL. CLASS. [26] | Dominant Species [25] | Typical Species [25] |
|---|---|---|---|---|
| 6120 Xeric sand calcareous grasslands | 6120_2 | 34.12 | *Poa angustifolia, Festuca ovina, Festuca rubra* | *Jasione montana, Hylotelephium* spp.*, Pilosella officinarum, Sedum acre, Thymus* spp.*, Veronica spicata, Viscaria vulgaris* |
| 6210 Semi-natural dry grasslands and scrubland facies on calcareous substrates | 6210_2, 6210_3 | 34.31 to 34.34 | 6210_2: *P. angustifolia, F. rubra, Fragaria vesca*<br>6210_3: *Helictotrichon pubescens, F. rubra, Fragaria viridis,* | 6210_2: *Agrimonia eupatoria, Carex caryophyllea, Centaurea scabiosa, Pimpinella saxifrage, Polygala comosa, Thymus ovatus*<br>6210_3: *Filipendula vulgaris, Medicato falcate, Plantago media, P. angustifolia, Polygala comosa, Potentilla reptans, Trifolium montanum* |
| 6270 Fennoscandian lowland species-rich dry to mesic grasslands | 6270_1, 6270_3 | 35.1212, 35.1223, 38.22, 38.241 | 6270_1: *Agrostis tenuis, Anthoxanthum odoratum, Briza media, Cynosurus cristatus, F. rubra*<br>6270_3: *Deschampsia caespitosa, F. rubra, Holcus lanatus* | 6270_1: *Alchemilla* spp.*, Dianthus deltoids, Leontodon hispidus, Leontodon autumnalis, P. media, Plantago lanceolate, Primula veris, Prunella vulgaris, Rhinanthus minor, Trifolium repens*<br>6270_3: *Filipendula ulmaria, Galium boreale, Geum rivale, Geranium palustre, Hierochloe odorata, Lychnis flos-cuculi, Scirpus sylvaticus, Carex cespitosa, Lysmachia nummularia* |
| 6410 Molinia meadows on calcareous, peaty, or clayey-silt-laden soils | 6410_4 | 37.31 | *Molinia caerulea, Festuca arundinacea, Filipendula ulmaria, H. pubescens, D. caespitosa* | *Carex buxbaumii, Carex flacca, Carex hartmanii, Carex hostiana, Carex panice, Galium boreale, Inula salicina, Polygala amarelle, Potentilla erecta, Scorzonera humilis, Succisa pratensis* |
| 6450 Northern boreal alluvial meadows | 6450_1 | - * | *Carex acuta, Carex acutiformis, Carex appropinguata, Carex elata, Carex paniculata, Carex vesicaria, Calamagrostis canescens, Phalaris arundinacea* | *Carex rostrata, Carex vulpina, Stellaria palustris, Lathyrus palustris, Lythrum salicaria, Veronica longifolia* |
| 6510 Lowland hay meadows | 6510_1 | 38.2 | *Arrhenatherum elatius, Bromopsis inermis, Festuca pratensis, H. pubescens* | *Crepis biennis, Heracleum sibiricum, Knautia arvensis, Pastinaca sativa, Tragopogon pratensis, Campanula patula, Centaurea jacea, Carum carvi, Galium album, Lathyrus pratensis* |

* Includes several vegetation types which vary according to the moisture (flooding) gradient: *C. acuta* or *C. aquatilis*-alluvial meadows, *Calamagrostis*-alluvial meadows, *Phalaris*-alluvial meadows, *Deschampsia caespitosa*-alluvial meadows.

Most of the samples (89) were collected in June–August of 2014. Thirty-nine and 34 samples were collected in 2015 and 2016, respectively (Table S1). Sampling in June (almost half of the samples) corresponded to a vegetation period when grassland biomass has the highest fodder value. August samples: period of late mowing.

The selection of semi-natural grassland sampling plot locations was based on visual assessment of the area. One most representative $1 \times 1$ m vegetation plot was selected and biomass was clipped at 2 cm above the ground level within the 1 x 1 m square using hand shears (Figure 1). First samplings were performed before the first cut or at the beginning of the grazing period (late June or early July). The second sample was collected in late July or August in sites managed by late mowing. In unmanaged sites, the third sample was also collected in September 2015 (9 samples in total). To evaluate the fermentable carbohydrate potential of early biomass, one sample from each habitat was collected in early June (season of 2015).

Prior to clipping, a description of the vegetation (vascular plant species richness) in each square was prepared. Then, the collected material was stored in pre-weighed plastic bags and brought to the laboratory for further analyses. If the biomass was not processed within one day, the samples were cut to fractions <20 cm, manually homogenized, and kept frozen (–18 °C) in sealable bags.

**Figure 1.** $1 \times 1$ m square frame sampling plots before (**A**) and after (**B**) collection of grass samples.

### 2.2. Dry Matter and Ash Content Analyses

A representative set of grass biomass was cut to pieces below 10 mm. Total dry weight (DW) was determined as weight after drying of sample at + 105 °C (laboratory oven 60/300 LSN, SNOL, Utena, Lithuania) for 24 h. Total ash content was measured according to a modified EN ISO 18122 [27]. In brief, the samples were heated at + 550 °C for 2.5 h (Laboratory furnace 8, 2/1100, SNOL). Volatile solid (VS) percentage was calculated as the difference between total dry mater and ash.

### 2.3. Enzymatic Hydrolysis

For enzymatic hydrolysis, a previously described method was used [23]. In brief, all biomass samples (fresh or frozen) were ground (Retsch, Grindomix GM200) to fractions below 0.5 cm. Then, 0.05 M sodium citrate buffer (mono–sodium citrate pure, AppliChem, Germany) was added to the biomass samples (final concentration, 9% w/v wet biomass) and mixed by vortexing. Then, the samples were boiled for 5 min (1 atm) to eliminate any indigenous microorganisms. After cooling to room temperature, a laboratory prepared enzyme (0.2 FPU/mL, obtained from white rot fungi *Irpex lacteus* (Fr.) Fr.) was added to the samples and incubated on an orbital shaker (New Brunswick, Innova 43) for 24 h at 30 °C and 150 rpm. Enzyme efficiency was compared with a commercial enzyme product (Viscozyme, Novozymes) and substrate control—hay (obtained in Latvia, 2015, DW 92.8 ± 1.3%).

Samples for reducing sugar measurements were collected after the addition of sodium citrate buffer, prior enzyme addition (both as zero time controls), and after 24 h of hydrolysis. All biomass samples were analyzed in six repetitions.

### 2.4. Reducing Sugar Analyses

The Dinitrosalicylic Acid (DNS) method was used to estimate the reducing sugar quantities in the collected samples [28]. First, the samples were centrifuged ($6600\times g$, 10 min). Then, 0.1 mL of the supernatant was mixed with 0.1 mL of 0.05 M sodium citrate buffer and 0.6 mL of DNS (SigmaAldrich, Taufkirchen, Germany). Distilled water was used as blank control. To obtain the characteristic color change, the samples were boiled for 5 min and transferred to cold water and supplied with 4 mL of distilled water. Absorption measurements were performed with a spectrophotometer (Camspec M501, Leeds, UK) at 540 nm. For absolute concentrations, a calibration curve against glucose was plotted.

### 2.5. Statistical Analyses

For data analysis, MS Excel 2013 *t*–test (two tailed distribution) and ANOVA single parameter tool (significance level ≤0.05) were used for analysis of variance on data from various sample setups.

## 3. Results and Discussion

### 3.1. Assessment of Biomass Resources

Biochemical parameters such as total solids (TS), volatile solids (VS), and ash content were analyzed for grass biomass samples collected from 6 habitats to evaluate the overall composition of the biomass and its changes over time. These parameters characterize the biomass as a potential energy source and indicate its absolute energetic value. Fast growing biomass can have ash content above 20%; woody biomass has typically 1% ash content. Each 1% increase in ash translates roughly into a decrease of 0.2 MJ/kg of heating value, making it an unpopular resource for combustion [29]. At the same time, the presence of inorganic chemicals can be a good source of microelements along with sugars in the fermentation processes.

The average dry matter from grassland samples in respective Community Importance habitats ranged roughly from 1.0 to 6.0 t/ha (Figure 2) and 93 ± 2% from the dry matter were volatile solids. The highest average yields were obtained from Lowland hay meadows (6510), but the lowest were from Xeric sand calcareous grasslands (6120). That corresponds to yields from semi-natural grasslands in Estonia [30], central Germany [31], and Denmark [32].

The harvesting time had a significant impact on the total amount of the biomass. On average, 5% to 32% less biomass was harvested in June than in July and 17.5 to 42.6 % less in June than in August.

Moreover, variations were observed among the harvesting years. The amount of the biomass (t/ha) in 2016 was 33% to 19% less than in 2015 and up to 27% less than in 2014 (Table 2). Assessment of average daily temperature did not present any significant fluctuations among the years (Figure S1). At the same time, total precipitation in both sampling locations during the summer months was lower in 2015 when compared to 2014 and 2016 (Figure S2). This, to some extent, could explain the differences between these years. A similar influence of annual weather conditions on yield in multi-species grassland has been reported from Estonia and Denmark [30,33].

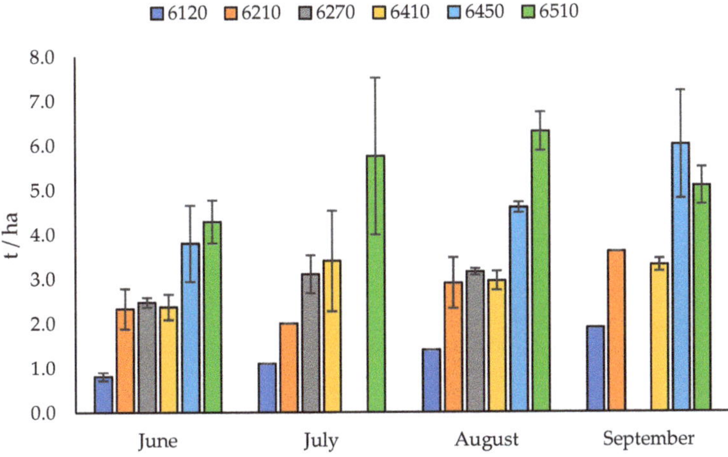


**Figure 2.** The average biomass dry matter (t/ha) collected from various grassland habitats at different sampling months over a three year period.

The ash content ranged from 3.84 to 9.62% from DW. The lowest ash content was observed in samples from Xeric sand calcareous grasslands (6120) (5.72 ± 1.03%) and the highest for semi-natural dry grasslands and scrubland facies on calcareous substrates (6210) (7.41 ± 1.10%, $p < 0.05$ among other biotopes). This corresponds to the results of other studies—the highest ash concentrations are typically identified in samples from the habitats with larger proportion of dicotyledonous plant species. Typically, ash content is associated with the concentration of minerals in plant organs [34] and dicotyledonous plants tend to accumulate greater quantities of minerals compared with monocotyledonous plants [33].

**Table 2.** The average quantity of biomass (t/ha as dry matter) collected from grassland habitats at various sampling years.

| Habitat Type | Average Dry Matter, t/ha | | |
|---|---|---|---|
| | 2014 | 2015 | 2016 |
| 6120 Xeric sand calcareous grasslands | 1.0 | 1.2 | 0.8 |
| 6210 Semi-natural dry grasslands and scrubland facies on calcareous substrates | 2.1 | 3.0 | 2.1 |
| 6270 Fennoscandian lowland species-rich dry to mesic grasslands | 2.8 | 3.2 | 2.6 |
| 6410 Molinia meadows on calcareous, peaty, or clayey-silt-laden soils | 3.0 | 2.9 | 2.2 |
| 6450 Northern boreal alluvial meadows | 4.5 | 5.1 | 3.5 |
| 6510 Lowland hay meadows | 4.4 | 5.7 | 3.9 |

*3.2. Enzyme Potential to Release Carbohydrates*

Prior to application of a non-commercial enzyme from *I. lacteus*, its efficiency to release fermentable sugars from hay was compared with a commercial enzyme product. The results demonstrated that a commercial preparation was able to release $0.39 \pm 0.05$ g/g hay DW after 24 h of incubation. Due to the variable species composition, the amount of cellulose and hemicellulose in hay can vary from 35–45% and 30–50%, respectively [35]. However, prolonged incubation (48 h) did not yield any significant increase ($p > 0.05$) and reached only $0.409 \pm 0.048$ g/g DW. At the same time, a crude non-commercial product (un-concentrated, un-purified) yielded $0.183 \pm 0.03$ g/g DW after 24 h and $0.199 \pm 0.045$ g/g DW after 48 h. In both cases, the amount of sugar released after mechanical and thermal pre-treatment was not significant. Despite lower yields ($p < 0.05$), the observed extractable sugar concentration was still higher than reported for various grass materials [36]. Due to lower costs and potential wide scale application, a non-commercial preparation was used for all future tests and 24 h incubation was set as the optimal.

*3.3. Fermentable Sugar Yields*

To evaluate the amount of fermentable sugar released from various grassland biomass resources, enzymatic hydrolysis with the non-commercial enzyme product at optimal conditions was performed. The results of 2014 showed significantly higher ($p < 0.05$) sugar yields (w/w) in June than in July or August (Table 3, Figure 3).

The length of the vegetation season had an overall tendency to decrease the amount of produced sugar. This was observed for all habitats in both 2014 and 2015 sampling seasons where June produced the highest sugar yields ($p < 0.05$) when compared to August or September. The samples from August and September demonstrated no significant sugar yield difference ($p > 0.05$).

Semi-natural dry grassland and scrubland facies on calcareous substrates (6210) and Lowland hay meadow (6510) samples produced the highest fermentable carbohydrate yields in 2014, e.g., 0.235 and 0.165 g per g VS, respectively. In 2015, the highest sugar yields were attributed to Xeric sand calcareous grasslands (6120) and 6210, but the lowest ones were in the samples of 6510 and Northern boreal alluvial meadows (6450) collected in September (Table 3). This slightly contradicted the results obtained in 2014, when from 6210, the highest yield (w/w) was obtained. One of the reasons for this could be the higher proportion of dicotyledonous plants in samples from 6210 collected during 2014. Similarly, as observed before, biomass with dominant monocotyledonous plant proportion showed lower carbohydrate yields due to higher crystallinity, lower hydrolysability, and potential presence of enzyme activity interfering substances [37].

The assessment of the overall producible sugar quantity from one ha exhibited a high potential of 6510 which from all tested habitats had the highest productivity in all vegetation periods, and in June, more than 0.7 t of fermentable sugar per ha could be produced. Other habitats that have demonstrated high sugar yields had lower productivity, e.g., 6210 having only 0.45 t/ha in June (Figure 3, Table 3) and 6120 even having below 0.2 t/ha.

In 2016, samplings were performed only in June with the aim to determine if there was any trend in-between habitats over the years. Again, the highest sugar yields (w/w) were produced from the habitat 6120, followed by Molinia meadows on calcareous, peaty, or clayey-silt-laden soils (6410) and 6510. Assessment of the total sugar quantity per 1 ha revealed that 6510 was able to generate more than 0.78 t of sugar per ha; however, 6120, only 0.186 t/ha. Similarly, as in previous seasons, this difference was due to the low total biomass quantity in 6120; thus, low correlation between fermentable sugar yield (per g biomass) and total amount of sugar per ha of habitat was observed.

The evaluation of the vegetation period showed a strong decrease in sugar yields with increasing vegetation time (Figure 3). No significant decrease ($p > 0.05$) was observed only between the samples collected in August and September. Similar observations have been made for methane yields in biogas production, where the increase in crude fiber at the end of the vegetation period has been set out as one of the main factors influencing the methane production [38]. Others have pointed out that to grasses harvested after October, an extra carbohydrate source must be added if applied for energy production purposes [36]. No influence of specific habitat type has been observed or recorded previously.

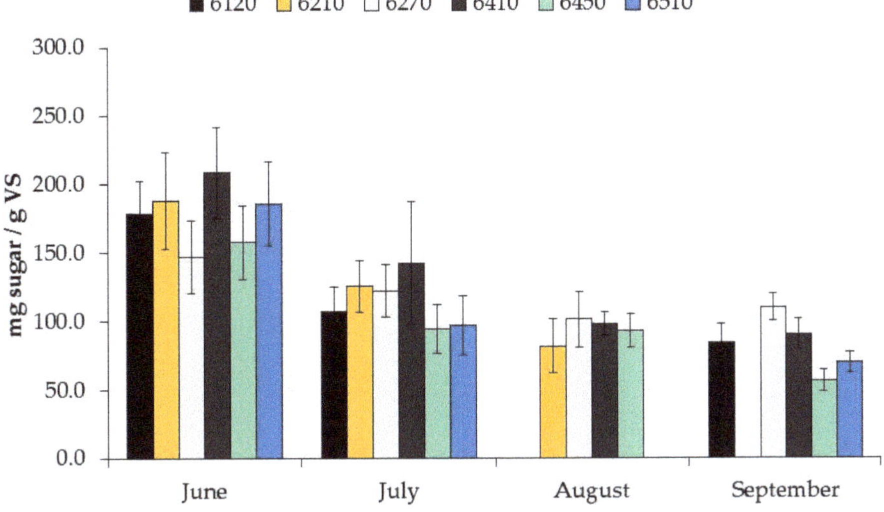


**Figure 3.** The amount of fermentable sugar produced per g volatile solid (VS) from biomass collected at various community importance habitats during 2014–2016 vegetation periods. Each bar represents the average value from at least two samplings with six individual measurements of reducing sugar.

**Table 3.** The Reducing sugar yield (mg/g volatile solid (VS) or t/ha) that can be produced from natural grassland habitats at various sampling period.

| EU Habitat Code | 2014 | | | | | | 2015 | | | | | | | 2016 | |
|---|---|---|---|---|---|---|---|---|---|---|---|---|---|---|---|
| | June | | July | | August | | June | | July | | August | | September | | June | |
| | mg/g VS | t/ha | mg/g VS | t/ha | mg/g VS | t/ha | mg/g VS | t/ha | mg/g VS | t/ha | mg/g VS | t/ha | mg/g VS | t/ha | mg/g VS | t/ha |
| 6120 | 147.61 ± 29.69 | 0.133 | 107.48 ± 23.29 | 0.118 | n/d | n/d | 225.46 ± 19.90 | 0.180 | n/d | n/d | n/d | n/d | 84.58 ± 13.56 | 0.161 | 233.35 ± 109.1 | 0.186 |
| 6210 | 235.49 ± 68.20 | 0.447 | 114.80 ± 24.75 | 0.230 | 81.71 ± 25.21 | 0.204 | 176.44 ± 22.80 | 0.493 | 158.00 ± 19.10 | n/d | n/d | n/d | n/d | n/d | 181.26 ± 28.61 | 0.380 |
| 6270 | n/d | n/d | 81.20 ± 24.88 | 0.227 | 101.34 ± 22.09 | 0.314 | n/d | n/d | 115.49 ± 31.47 | 0.393 | n/d | n/d | 109.87 ± 10.01 | n/d | 147.06 ± 26.92 | 0.382 |
| 6410 | n/d | n/d | 88.49 ± 13.33 | 0.265 | 103.56 ± 6.64 | 0.321 | 139.66 ± 6.40 | 0.377 | 142.48 ± 44.60 | 0.369 | 97.87 ± 11.77 | 0.274 | 90.23 ± 1.68 | 0.298 | 203.67 ± 50.59 | 0.447 |
| 6450 | 157.08 ± 46.71 | 0.659 | 94.10 ± 3.94 | 0.489 | n/d | n/d | 152.32 ± 9.38 | 0.669 | n/d | n/d | 92.84 ± 11.9 | 0.427 | 56.16 ± 18.3 | 0.337 | 161.98 ± 37.01 | 0.564 |
| 6510 | 164.74 ± 50.59 | 0.725 | 90.55 ± 25.80 | 0.498 | n/d | n/d | 166.66 ± 5.44 | 0.783 | 105.48 ± 15.17 | 0.738 | n/d | n/d | 69.71 ± 4.83 | 0.356 | 201.88 ± 36.01 | 0.784 |

n/d—not determined; VS—volatile solids.

In some cases, discrepancies from general observations have been detected. Molinia meadows on calcareous, peaty, or clayey-silt-laden soils (6410) did not produce the observed decrease in sugar yields with the progression of the vegetation season. This could be linked to the fact that 6410 includes Molinion grasslands, grasslands with low height sedge species like *Carex flacca*, *Carex hartmanii*, *Carex hostiana*, *Carex panicea*, *Carex buxbaumii*, as well as grasslands lacking any predominant species. Usually these habitats are represented with high species diversity and located in periodically drying soils [25]. One of the possible explanations can be related to the fact that in July 2014 and June 2015, the samples were collected mainly in sedge grasslands, while in August 2014 and July 2015, in Molinia grasslands. Furthermore, both sugar yield and productivity in 6270 was higher in August 2014 than in July—0.081 and 0.101 g/g VS or 0.22 and 0.31 t/ha, respectively. Apart from the general view (the increase in biomass and carbohydrate yields progresses with the vegetation time) that is challenged within this study, we hypothesize that the observed trend in 6270 is more linked to the environmental conditions, species composition in each individual sampling plot, and vegetation structure in general. Even in one habitat, multiple subtypes with diverse plant communities can be found. Nevertheless, to give the precise explanations of these variations, a more sophisticated classification and evaluation of species compositions would be required.

The average amount of the fermentable sugars highly varied not only seasonally, but also among the years. Sugar yields from the biomass harvested in June 2016 (a month with the most comprehensive data set) were 3% to 58% higher than in those collected in June 2014 and June 2015 for all habitats except 6210 (Table 3). Furthermore, it was estimated that the sugar yields tend to fluctuate ($p < 0.05$) even on a monthly basis, e.g., samples collected within the first ten days of June and at the end of June. The rationale for these differences within one habitat can be explained by the habitat's heterogeneity. The habitats listed in the annexes of the EU Habitats Directive are not classified in a single hierarchical system. Habitats can be separated by the phytosociological classification of plant communities or by habitat groups that include several similar habitats. These can be further divided by specific environmental conditions. Moreover, weather conditions could affect the productivity in single habitat on a yearly basis.

The management of natural grasslands in Natura 2000 classified territories is generally restricted to low-intensity agricultural practices and strict regulations related to grazing, mowing, and cutting [9]. Despite grazing being seen as one of the simplest strategies, follow up on over- or under-grazing, formation of patchiness, preference of certain species by animals, or maintenance of cattle are limiting factors. Mowing at the same time requires the selection of correct timing and frequency; e.g., late moving is preferred to protect animal species and late-flowering plants. At the same time, early cutting and removal of cut grass help to maintain low nutrient levels, keep plant diversity, and avoid alien species [9,39].

On average, the amount of sugar produced from the various grassland habitats at various vegetation periods was comparable to the data obtained with hay (~0.2 g/g DW) and the strategy was shown to be applicable in both high productivity grasslands and at early cutting periods. Upgraded enzymes, adjustment of the technology, e.g., introduction of more intense pre-treatment, could further facilitate the release of the energy stored into grassland biomass. Nevertheless, as demonstrated by this study, multispecies presence, quantities, and applicability under variable conditions set grassland resources as highly sustainable when fermentable carbohydrate production is foreseen.

## 4. Conclusions

A simple pre-treatment/hydrolysis technique with non-commercial enzymes made from *I. lacteus* was demonstrated to be efficient for the production of fermentable sugars from the biomass of community important grassland habitats classified under Natura 2000 that have to follow restricted farming practices.

The results showed that fermentable sugar yields from semi-natural grassland habitats are closely linked to vegetation period and plant species variation (monocotyledonous/dicotyledonous species proportion). Dicotyledonous plant rich habitats (6120, 6210) at the beginning of vegetation generated the highest amount of fermentable sugar per mass of biomass—above 0.2 g per g VS. At the same time, habitats rich in total biomass (6510) yielded higher sugar quantities per ha. The lowest yield and productivity in all habitats were observed in August–September, indicating potential bottlenecks of bioenergy production when biomass is collected at a late vegetation period. Overall, the study demonstrated that fermentable carbohydrate production from multispecies biomass of natural and semi-natural grasslands can be used as an alternative management strategy to currently practiced grazing. Thus, fuel production technologies can be merged with sustainable environment management.

**Supplementary Materials:** The following are available online at https://www.mdpi.com/1996-1073/14/5/1312/s1, Annex 1: Location of biomass sampling plots, Table S1: Number of collected biomass samples per sampling year and habitat type; Figure S1: Average daily temperature in sampling months of 2014, 2015 and 2016 at 2 locations; Figure S2: Total precipitation (mm) in sampling months of 2014, 2015 and 2016 at 2 locations and the whole period (Total).

**Author Contributions:** Conceptualization, writing, and data analysis, L.M.; validation, T.J.; formal analysis and data collection, B.D., E.S.; sampling, B.S. All authors have read and agreed to the published version of the manuscript.

**Funding:** The work was supported by the IPP3: INNO INDIGO Programme Project B-LIQ 'Development of an Integrated Process for Conversion of Biomass to Affordable Liquid Biofuel', No. ES/RTD/2017/18, and the National Research Programme "Energetics" Project "Innovative solutions and recommendations for increasing the acquisition of local and renewable energy resources in Latvia" No. VPP-EMAER-2018/3-0004.

**Institutional Review Board Statement:** Not applicable.

**Informed Consent Statement:** Not applicable.

**Data Availability Statement:** Not applicable.

**Acknowledgments:** We acknowledge the EU LIFE+ Nature & Biodiversity program Project "GRASSSERVICE"—Alternative use of biomass for maintenance of grassland biodiversity and ecosystem services (LIFE12 BIO/LV/001130) for access to biomass samples and data on sampling plots.

**Conflicts of Interest:** The authors declare no conflict of interest.

## References


1.  European Commission; Directive (EU). 2018/2001 on the promotion of the use of energy from renewable sources. *Off. J. Eur. Union* **2018**, *5*, 82–209.
2.  Basile, A.; Dalena, F. *Second and Third Generation of Feedstocks: The Evolution of Biofuels*; Elsevier: Amsterdam, The Netherlands, 2019; pp. 213–240.
3.  Pandey, A.; Negi, S.; Binod, P.; Larroche, C. *Pretreatment of Biomass*, 1st ed.; Elsevier: Amsterdam, Netherlands, 2015; pp. 1–272.
4.  Jungers, J.M.; Fargione, J.E.; Sheaffer, C.C.; Wyse, D.L.; Lehman, C. Energy potential of biomass from conservation grasslands in Minnesota, USA. *PLoS ONE* **2013**, *8*, e61209. [CrossRef]
5.  European Commission. 15 December 2020 Eurostat. Utilized Agricultural Area by Categories. Available online: https://ec.europa.eu/eurostat/databrowser/view/tag00025/default/table?lang=en (accessed on 30 December 2020).
6.  Wilson, J.B.; Peet, R.K.; Dengler, J.; Partel, M. Plant species richness: The world records. *J. Veg. Sci.* **2012**, *23*, 796–802. [CrossRef]
7.  Isselstein, J.; Jeangros, B.; Pavlu, V. Agronomic aspects of biodiversity targeted management of temperate grasslands in Europe—A review. *Agron. Res.* **2005**, *3*, 139–151.
8.  Ostermann, O.P. The need for management of nature conservation sites designated under Natura 2000. *J. Appl. Ecol.* **1998**, *35*, 968–973. [CrossRef]
9.  European Commission. *Farming for Natura 2000. Guidance on How to Support Natura 2000 Farming Systems to Achieve Conservation Objectives, Based on Member States Good Practice Experiences*; Publications Office of the European Union: Luxembourg, 2018; pp. 1–146.
10. Poschlod, P.; Bakker, J.P.; Kahmen, S. Changing land use and its impact on biodiversity. *Basic Appl. Ecol.* **2005**, *6*, 93–98. [CrossRef]
11. Dengler, J.; Janisová, M.; Török, P.; Wellstein, C. Biodiversity of Palaearctic grasslands: A synthesis. *Agric. Ecosyst. Environ.* **2014**, *182*, 1–14. [CrossRef]
12. Bobbink, R.; Den Dubbelden, K.; Willems, J.H. Seasonal dynamics of phytomass and nutrients in chalk grassland. *Oikos* **1989**, *55*, 216–224. [CrossRef]
13. Spehn, E.M.; Scherer-Lorenzen, M.; Schmid, B.; Hector, A.; Caldeira, M.C.; Dimitrakopoulos, P.G.; Finn, J.A.; Jumpponen, A.; O'Donnovan, G.; Pereira, J.S.; et al. The role of legumes as a component of biodiversity in a cross European study of grassland biomass nitrogen. *Oikos* **2002**, *98*, 205–218. [CrossRef]
14. Poptcheva, K.; Schwartze, P.; Vogel, A.; Kleinebecker, T.; Hölzel, N. Changes in wet meadow vegetation after 20 years of different management in a field experiment (North-West Germany). *Agric. Ecosyst. Environ.* **2009**, *134*, 108–114. [CrossRef]
15. Neuenkamp, L.; Metsoja, J.-A.; Zobel, M.; Hölzel, N. Impact of management on biodiversity-biomass relations in Estonian flooded meadows. *Plant Ecol.* **2013**, *214*, 845–856. [CrossRef]
16. Janišová, M.; Michalcová, D.; Bacaro, G.; Ghisla, A. Landscape effects on diversity of semi-natural grasslands. *Agric. Ecosyst. Environ.* **2014**, *182*, 47–58. [CrossRef]
17. Cousins, S.A.O.; Eriksson, O. The influence of management history and habitat on plant species richness in a rural hemiboreal landscape, Sweden. *Landsc. Ecol.* **2002**, *17*, 517–529. [CrossRef]
18. Caspeta, L.; Buijs, N.A.A.; Nielsen, J. The role of biofuels in the future energy supply. *Energy Environ. Sci.* **2013**, *6*, 1077–1082. [CrossRef]
19. Tsapekos, P.; Khoshnevisan, B.; Alvarado-Morales, M.; Symeonidis, A.; Kougias, P.G.; Angelidaki, I. Environmental impacts of biogas production from grass: Role of co-digestion and pretreatment at harvesting time. *Appl. Energy* **2019**, *252*, 113467. [CrossRef]
20. Leclere, D.; Valin, H.; Frank, S.; Havlik, P. *Assessing the Land Use Change Impacts of Using EU Grassland for Biofuel Production. Task 4b of Tender ENER/C1/2013-412*; ECOFYS Netherland B.V.: Utrecht, The Netherlands, 2016; pp. 1–49.
21. Xu, N.; Liu, S.; Xin, F.; Zhou, J.; Jia, H.; Xu, J.; Jiang, M.; Dong, W. Biomethane production from lignocellulose: Biomass Recalcitrance and its impacts on anaerobic digestion. *Front. Bioeng. Biotechnol.* **2019**, *7*, 1–12. [CrossRef] [PubMed]
22. Sawatdeenarunat, C.; Surendra, K.C.; Takara, D.; Oechsner, H.; Khanal, S.K. Anaerobic digestion of lignocellulosic biomass: Challenges and opportunities. *Bioresour. Technol.* **2015**, *178*, 178–186. [CrossRef]
23. Mezule, L.; Berzina, I.; Strods, M. The impact of substrate-enzyme proportion for efficient hydrolysis of hay. *Energies* **2019**, *12*, 3526. [CrossRef]
24. Anonymous. Interpretation Manual of European Union Habitats. European Commission. DG Environment. 28 April 2013. Available online: https://ec.europa.eu/environment/nature/legislation/habitatsdirective/docs/Int_Manual_EU28.pdf (accessed on 6 February 2021).
25. Aunins, A.; Aunina, L.; Bambe, B.; Engele, L.; Ikauniece, S.; Kabucis, I.; Laime, B.; Larmanis, V.; Reriha, I.; Rove, I.; et al. *Eiropas Savienības Aizsargājamie Biotope Latvijā. Noteikšanas Rokasgrāmata*; Latvijas Dabas Fonds: Riga, Latvia, 2013; pp. 1–391. (In Latvian)
26. Devillers, P.; Devillers-Terschuren, J. A classification of Palaearctic habitats. *Nat. Environ.* **1996**, *78*, 1–157.
27. EN ISO 18122:2016. *Solid biofuels—Determination of Ash Content*; International Organization for Standardization: Geneva, Switzerland, 2015; pp. 1–6.
28. Ghose, T.K. Measurement of cellulose activities. *Pure Appl. Chem.* **1987**, *59*, 257–268. [CrossRef]
29. Jenkins, B.M.; Baxter, L.L.; Miles, T.R., Jr.; Miles, T.R. Combustion properties of biomass. *Fuel Process. Technol.* **1998**, *54*, 17–46. [CrossRef]

30. Melts, I. Biomass from semi-natural grasslands for bioenergy. Ph.D. Thesis, Estonian University of Life Sciences, Tartu, Estonia, 2014; p. 125.
31. Wachendorf, M.; Richter, F.; Fricke, T.; Graß, R.; Neff, R. Utilization of semi-natural grassland through integrated generation of solid fuel and biogas from biomass. I. Effects of hydrothermal conditioning and mechanical dehydration on mass flows of organic and mineral plant compounds, and nutrient balances. *Grass Forage Sci.* **2009**, *64*, 132–143. [CrossRef]
32. Hensgen, F.; Buhle, L.; Donnison, I.; Heinsoo, K.; Watchendorf, M. Energetic conversion of European semi-natural grassland silages through the integrated generation of solid fuel and biogas from biomass: Energy yields and the fate of organic compounds. *Biores. Technol.* **2014**, *154*, 192–200. [CrossRef] [PubMed]
33. Pirhofer-Walzl, K.; Søegaard, K.; Høgh-Jensen, H.; Eriksen, J.; Sanderson, M.A.; Rasmussen, J.; Rasmussen, J. Forage herbs improve mineral composition of grassland herbage. *Grass Forage Sci.* **2011**, *66*, 415–423. [CrossRef]
34. Monti, A.; Di Virgilio, N.; Venturi, G. Mineral composition and ash content of six major energy crops. *Biomass Bioenergy* **2008**, *32*, 216–223. [CrossRef]
35. Chen, Y.; Sharma-Shivappa, R.R.; Keshwani, D.; Chen, C. Potential of agricultural residues and hay for bioethanol production. *Appl. Biochem. Biotechnol. Part A Enzyme Eng. Biotechnol.* **2007**, *142*, 276–290. [CrossRef] [PubMed]
36. Herrmann, C.; Prochnow, A.; Heiermann, M.; Idler, C. Biomass from landscape management of grassland used for biogas production: Effects of harvest date and silage additives on feedstock quality and methane yield. *Grass Forage Sci.* **2013**, *69*, 549–566. [CrossRef]
37. Zoghlami, A.; Paës, G. Lignocellulosic biomass: Understanding recalcitrance and predicting hydrolysis. *Front. Chem.* **2019**, *18*, 1–11. [CrossRef]
38. Prochnow, A.; Heiermann, M.; Plöchl, M.; Linke, B.; Idler, C.; Amon, T.; Hobbs, P.J. Bioenergy from permanent grassland—A review: 1. Biogas. *Bioresour. Technol.* **2009**, *100*, 4931–4944. [CrossRef]
39. Calaciura, B.; Spinelli, O. Management of Natura 2000 Habitats. 6210 Semi-Natural Dry Grasslands and Scrubland Facies on Calcareous Substrates (*Festuco-Brometalia*). European Commission. 2008. Available online: https://ec.europa.eu/environment/nature/natura2000/management/habitats/pdf/6210_Seminatural_dry_grasslands.pdf (accessed on 16 February 2021).

*Article*

# Repurposing Fly Ash Derived from Biomass Combustion in Fluidized Bed Boilers in Large Energy Power Plants as a Mineral Soil Amendment

Elżbieta Jarosz-Krzemińska [1,*] and Joanna Poluszyńska [2]

[1] Department of Environmental Protection, Faculty of Geology, Geophysics and Environmental Protection, AGH University of Science and Technology, Al. A. Mickiewicza 30, 30-059 Krakow, Poland

[2] Łukasiewicz Research Network—Institute of Ceramics and Building Materials, Oświęcimska Street 21, 45-641 Opole, Poland; j.poluszynska@icimb.pl

* Correspondence: ekrzeminska@agh.edu.pl

Received: 10 August 2020; Accepted: 11 September 2020; Published: 14 September 2020

**Abstract:** This research involved studying the physico-chemical parameters of fly ash derived from the combustion of 100% biomass in bubbling and circulating fluidized bed boilers of two large energy plants in Poland. Chemical composition revealed that ash contains substantial amounts of CaO (12.86–26.5%); $K_2O$ (6.2–8.25%); MgO (2.97–4.06%); $P_2O_5$ (2–4.63%); S (1.6–1.83%); and micronutrients such as Mn, Zn, Cu, and Co. The ash from the bubbling fluidized bed (BFB) was richer in potassium, phosphorus, CaO, and micronutrients than the ash from the circulating fluidized bed (CFB) and contained cumulatively less contaminants. However, the BFB ash exceeded the threshold values of Cd to be considered as a liming amendment. Additionally, according to our European Community Bureau of Reference (BCR) study Pb and Cd were more mobile in the BFB than in the CFB ash. Except for a low nitrogen content, the ash met the minimum requirements for mineral fertilizers. Acute phytotoxicity revealed no inhibition of the germination and seed growth of *Avena sativa* L. and *Lepidium sativum* plants amended with biomass ash. Despite the fact that low nitrogen content excludes the use of biomass fly ash as a sole mineral fertilizer, it still possesses other favorable properties (a high content of CaO and macronutrients), which warrants further investigation into its potential utilization.

**Keywords:** fly ash; biomass combustion; fluidized bed boilers; acute phytotoxicity test; mineral fertilizer; BCR sequential extraction; metal speciation

## 1. Introduction

Globally, almost one third of electricity is generated from coal; despite this fact, renewable energy sources such as biomass are increasingly gaining a foothold. Twenty-eight European Union (EU) countries are obliged to meet certain targets regarding their share of energy from renewable sources in gross energy production by the year 2020 (according to the EU Directive 2009/28/WE). In 2020, this target is 20% for most EU countries, whereas Poland has to meet a target of 15%. Consequently, the European Environment Agency has indicated that the use of biomass in large combustion plants in the EU has tripled between years 2004 and 2016. For instance, in Poland almost half of the electricity derived from renewable sources comes from biomass. Investments in energy generation derived from biomass are either in the planning stage or have already been implemented in many of Poland's heat and electric power plants.

This investment "boom" has resulted in the generation of an entirely new type of waste. The resultant by-products derived from the combustion of 100% biomass in large power plants as well as power and heat installations are very different from the biomass ash derived from their smaller counterparts. Compared to conventional fly ash from coal combustion, biomass fly ash has a different composition as well as its own unique characteristics and properties. Therefore, it needs to be stressed that the term "fly ash" should not be regarded as a universal term, since the type of combustion technology as well as type of combustion feedstock (coal, biomass, or biomass co-combustion) generates different types of fly ash. According to the Polish Waste Catalog [1], biomass combustion by-products originating from fluidized bed boilers are classified in the same group as the conventional fly ash from coal combustion (10 01 82). Consequently, this type of waste also undergoes a different utilization pathway than that of other biomass combustion by-products such as fly ash, originating from peat and untreated wood (waste code 10 01 03), or the waste from combusting straw in municipal boilers (10 01 99).

Until now, conventional fly ash, derived from power plants fired with coal fuel, was commonly used for the production of building materials, general-purpose cements, building ceramics, hydraulic binders, and binding materials, and in road construction (road base). The fly ash generated from biomass combustion, due to its unfavorable composition, is not suitable for traditional management methods. For instance, it contains a substantial amount of phosphorus, which slows hydration and extends the setting time of concrete, thus causing a reduction in its strength. However, while this chemical parameter precludes the use of biomass ash in construction materials, still it is a highly desirable attribute in a different utilization direction, such as as a potential mineral fertilizer or soil improver in land use. Fluidized bed boilers are the most commonly recommended type of boiler for combusting biomass fuel, especially in the process of heat and energy production from biomass in large and very large combustion plants. The energy sector uses either bubbling fluidized bed (BFB) technology or its upgraded version, circulating fluidized bed (CFB), sometimes called second-generation boilers. Both combustion technologies are characterized by very high thermal efficiencies of up to 87%, however circulated fluidized bed furnaces are more commonly applied in larger scale power installations. As indicated by Pallarès and Johnsson [2], in BFB technology (referred also as stationary fluidized bed) the combustion mostly takes place in the bed and in the lower part of the freeboard, and there is no external recirculation of the bed, unlike that of CFB boilers, which operate under circulating conditions where, unlike in bubbling beds, combustion is distributed more homogeneously along the height of the furnace. CFB employs a higher gas velocity [3] and/or finer bed solids than those used in BFB.

Biomass combustion by-products are fly ash that is captured by electrostatic precipitators, as well as bottom ash that is collected directly from the grate. Since, in fluidized bed furnaces, sorbents such as ground limestone, dolomite, or lime are used to bind sulfur compounds and control $SO_2$ emission, the solid residue also contains substantial amounts of desulfurization products, such as calcium sulfate [4]. In order to ensure optimal sulfur binding conditions, the temperature in the furnace chamber is maintained at a level of 850 to 900 °C.

As previously stated, biomass combustion by-products are a very heterogenous group of waste materials, whose chemical, mineralogical, and physical characteristics vary significantly between installations. Its final chemical composition is influenced by a multitude factors—i.e., the biomass source and origin, the energy plant's age, the harvesting time, the proportion of biomass/feedstock mixture, the soil and biomass growing conditions, the combustion temperature, the type of sorbent used in the combustion process, or even its granulometry and many other factors [5,6]. Theoretically, the same technological process of combustion and the same feedstock used but delivered from a different source may influence the composition of the final by-product ash. For that reason, it is especially important to characterize each type of biomass ash individually prior to finding an appropriate utilization approach.

Regardless of the occurrence of the great variability of biomass ash—as confirmed, e.g., by Vassilev et al. [6] in their review of almost 600 articles on the topic—the vast majority of researchers agree on the fact that the prevailing types of biomass fly ash derived from both small and laboratory installations [7–9], as well as those from bigger installations such as large or very large fluidized bed boilers [10–13], have good nutritional properties. Researchers report that [14–19], except for its low N content (which gets volatized during the combustion process [19]), biomass combustion by-products' chemical composition is similar to that of mineral fertilizers. Despite the fact that this type of waste is nearly free of nitrogen, it contains substantial amounts of other micro (B, Fe, Mo, Mn, Cu, Zn, Co) and macronutrients, both primary (P, K) and secondary (S, Mg, Ca, Na) [20], which are highly favorable in maintaining appropriate conditions conducive for plant growth. Ohenoja et al. [21], for instance, in their broad study on the utilization potential of biomass fly ash where they reviewed at least 46 research papers, confirmed the low content of contaminates in most ash derived from the fluidized bed combustion of pure biomass and satisfactory levels of macronutrients such as phosphorus and calcium, thus concluding the high feasibility of using it as a soil amendment. Moreover, the agronomic effects of using biomass fly ash derived from a variety of feedstocks and combusting technologies on crop yields were also reported by several authors [22–25]. For instance, the P fertilization effects of various biomass ash, such as rape meal ash, cereal ash, or straw ash, on eight types of crops (e.g., as maize, lupin, summer barley, oilseed rape, oil radish, etc.) were evaluated during pot experiments conducted by Schiemenz and Lobermann [15]. The authors concluded that the above-mentioned biomass ash can be an adequate source of phosphorus, even comparable with highly soluble commercial P fertilizers. Furthermore, Meller and Bilenda's findings [26] also confirm the fertilizing potential of biomass fly ash originating from BFB boilers in heat plants, which combust wood and agricultural feedstock. In their in situ experiment conducted on *Miscanthus sacchariflorus* grown on soil fertilized with the addition of BFB fly ash, the authors confirmed that an increased dose of ash caused a significant increase in the amount of available potassium, phosphorus, and magnesium in the soil. They reported that, as a result of amending the soil with 10.5 Mg·ha$^{-1}$ of BFB, the bioavailable phosphorus content in soil was increased by 27.06 mg/100 g, the bioavailable magnesium content by 15.05 mg/100 g, as well as bioavailable potassium by as much as 74.04 mg/100 g. In other research conducted by Ayeni et al. [27], reports of the positive effect of sawdust and wood ash applications on the enhancement of the N and P nutrient content as well as on growth of cocoa seedlings were presented.

In general, the utilization of biomass ash as a soil amendment or fertilizer (field or forest fertilizer) has a long history, especially in Nordic countries. As indicated by Ohenoja [21], Finland, for instance, is an undisputed leader in using this type of waste as a soil amendment. As a result, most research on the utilization potential of biomass ash is also carried out there [20,28–33]. It is a normal utilization practice in Finland to use biomass ash as a sole field or forest fertilizer when it is pretreated (e.g., granulated) and meets the threshold values set for both the contaminant and nutrient content. Both Finland and Denmark have established national legislation dedicated exclusively to ash recycling and fertilizing in forestry [34,35], with set threshold values for the total concentration of detrimental contaminants in ash (As, Cd, Cr, Cu, Ni, Pb, and Zn), as well as the minimum content of nutrients (Ca%, K + K%) required for both field and forest fertilizers. Moreover, the EU theoretically favors the application of biomass ash in top soil, since it fits well into the circular economy approach, however it should be noted that, at the same time, other EU legislation regarding the protection of top soil basically excludes this type of waste for land application, because of the strict limits regarding heavy metal content, which, in some cases, should be considered more as micronutrients than as contaminants. Van Dijen et al. [36] even concludes that EU policies regarding the utilization of biomass ash in agricultural and forest use are contradictory. However, in 2019 the EU recently revised previous fertilizer regulations and delivered a new regulation, (EU) 1009/2019 [37], which will take effect starting from 16 July 2022. This legislation repealing the "old" (EC) No. 2003/2003 regulation [38] will allow and support the general idea of using organic, bio-waste, or recycled fertilizers, such as biomass ash, in top soil as a fertilizer, liming agent, or soil improver alone and in addition to other fertilizing products if the waste has met new limits and threshold values. What is especially important and exceptional in this legislation is that individual EU countries can still set their own national fertilizing legislation with less strict limits, and the EU will still allow these non-conforming products to be available on the market, though not exported as CE products. Consequently, if biomass ash will be suitable and designated for the purpose of top soil application, it will lose its status as a waste. These are key changes that may act as a springboard, leading to the broader use of this by-product as a soil improver, fertilizer, or liming agent.

The aim of this study was to determine the chemical composition of fly ash resulting from the combusting of 100% biomass in two different types of fluidized bed boiler collected from two very large power plants in Poland (of a capacity of 183 MW and 205 MW), with particular emphasis placed upon the fertilizing properties of ash, as well as content of micro and macronutrients. Moreover, in order to determine the potential toxicity and bioavailability of elements to plants, an aqueous leaching test followed by a three-step sequential extraction European Community Bureau of Reference (BCR) was performed. Furthermore, acute phytotoxicity tests were conducted in order to evaluate the potential influence of the amendment of biomass fly on soil on select plant growth by determining the inhibition of seed germination, the inhibition of root elongation, and ultimately by calculating their germination index. This is the first part of broader research focused on finding the most suitable utilization approach for fly ash from biomass combustion in fluidized bed boilers generated by the Polish large-scale energy sector in millions of tonnes per annum.

## 2. Materials and Methods

### 2.1. Materials

The material in this research consisted of fly ash derived from the combustion of 100% biomass in the fluidized bed boilers of two very large energy plants in Poland. Fluidized bed boilers are the combustion technology employed in both plants.

The first power plant (for the purpose of this research called installation "BFB") is equipped with Poland's biggest bubbling fluidized bed boiler (BFB), which has a capacity of 183 MW, whereas the second power plant (for the purpose of this research called installation "CFB") in turn uses Poland's biggest circulating fluidized bed boiler (CFB), with a capacity of 205 MW. Both power plants are referred to as "green" installations, since they combust 100% biomass composed of a mixture of wood and agricultural residue ("agro") feedstock. The addition of "agro" biomass to the total weight of biomass feedstock is mandatory in Poland's energy production units, which are greater than 5 MW, and its share is strictly regulated. This mandatory inclusion of "agro" biomass (e.g., sunflower husk, different agricultural residues, energy crops, etc.) to the combustion process is quite troublesome. This issues arise from the fact that it has different physico-chemical properties than forest biomass, including particularly high levels of chlorine; sulfur; and alkali metals such as phosphorus, potassium, and sodium, all of which can cause corrosion and other technical problems. All our fly ash samples were collected in 2016/2017, when the minimum addition of "agro" biomass for the purpose of combustion in large "green" power units in Poland was set at a minimum of 20%.

The biomass feedstock in the CFB plant included 80% wood pellets, and the remaining 20% was agricultural waste, which consisted of sunflower husks. The biomass feedstock for the BFB installation consisted of 79% wood pellets, and the remaining 21% was agricultural residue (18% sunflower husks and 3% straw pellets). For further analysis, a total of four samples of fly ash were collected from the electrostatic precipitator, two from each installation. Samples of 20 kg each were homogenized, averaged, and determined to be representative for further analysis and tests.

### 2.2. Methods

The chemical composition of all fly ash samples was determined via X-ray fluorescence, using a WD-XRF ZSX Primus II Rigaku Spectrometer. Qualitative spectrum analysis was performed by identifying spectral lines, determining their possible coincidences, and then selecting analytical lines. The semi-quantitative analysis was conducted using the SQX Calculation program (fundamental parameter method), and was carried out in ranges from fluorine to uranium (F-U). Furthermore, the content of the determined elements was then normalized to 100%. Prior XRF analysis samples were prepared using a standard pelleting technique with the addition of a binder (Celleox) in a 4:2 proportion.

The concentration of chloride in ash was additionally determined using the titration method according to the European standard EN 196-2:2013 [39]. In order to determine the total concentration of metals in fly ash, samples were extracted with conc. nitric acid and hydrogen peroxide using microwave oven PRO, Anton PAAR, following digestion protocol PN-EN 13,657:2006 [40]. Furthermore, the concentrations of metals V, Cr, Mn, Co, As, Cd, Sn, Sb, Tl, and Pb where then analyzed via ICP-MS (Agilent 7700x) according to PN-EN ISO 17294-2:2016-11 [41]. The detection limits for particular elements in the ICP-MS apparatus were 0.25–2500 mg/kg. Mercury in raw samples was measured using atomic absorption spectrometry with amalgamation (AMA 254), according to an Ł-ICIMB accredited procedure: PB-LL-10 ed. 2 of 04/09/2017. The detection limits for this analytical device ranged from 0.005 to 100 mg/kg. The nitrogen and sulfur content was determined using the elemental analyzer CHNS + Cl + O Vario MACRO Cube by Elementar, using a high-temperature combustion method with TCD detection. Measurements were conducted according to PN-EN 15407:2011 [42] for N and PN-EN 15408:2011 [43] for S. The detection limits of the elemental analyzer device for N were 0.05–10% and were 0.1–8% for S. The primary nutrients in fly ash (P, N, K, S, Mg, Na, Ca) were expressed in both elemental as well as oxide forms, as requested by the EU fertilizer legislation act (EC) No. 2003/2003 [38], using the following conversion values: phosphorus (P) = phosphorus pentoxide ($P_2O_5$) × 0.36; potassium (K) = potassium oxide ($K_2O$) × 0.830; calcium (Ca) = calcium oxide (CaO) × 0.715; magnesium (Mg) = magnesium oxide (MgO) × 0.603; sodium (Na) = sodium oxide ($Na_2O$) × 0.742; (d) sulfur (S) = sulfur trioxide ($SO_3$) × 0.400.

Furthermore, a 24 h aqueous leaching test was conducted in order to determine the toxicity and thus potential mobility and bioavailability of elements in fly ash, according to PN-EN 12457-2006 [44]. This simple one-step test consisted of leaching ash for 24 h with distilled water in the ratio of 10:1 water to dry weight of the sample. The concentrations of chloride, sulfate, nitrate, and phosphate anions in the leachates were then determined according to PN-EN_ISO 10304-1:2009/AC 2012 [45], and the concentrations of sodium, potassium, calcium, and magnesium cations were detected according to PN EN ISO 14911:2002 [46] using the ion chromatography method (Metrohm IC 850 Professional with a conductometric detector and UVVIS). The concentration of the "leachable" and easily soluble metals, such as V, Cr, Mn, Co, Ni, Cu, Zn, As, Cd, Sn, Tl, and Pb, were then determined using the ICP-MS method. A more appropriate study on the bioavailability of primary nutrients—that is, potassium and phosphorus—was then conducted using the calorimetric method ($P_2O_5$), as well as the flame photometric method ($K_2O$). The available phosphorus content in ash was determined in accordance with PN-R-04023:1996 [47] using the MERCK SQ118 calorimeter. The content of bioavailable potassium was determined in accordance with the PN-R-04022: 1996/A1: 2002 [48] using the Zeiss flame photometer.

A speciation study of all the fly ash was performed using the 3-step sequential extraction proposed by the European Community Bureau of Reference (BCR), and delivered as a standardized and improved method of sequential extraction, mainly the commonly used 6th-step extraction according to Tessier et al. [49].

The dried ash samples (1 g, 2 h at 105 °C) were subjected to 3-step extraction according to the procedure provided by Ure et al. [50], included in Table 1. Samples were subjected to each extraction step using solutions of increasing aggressiveness in order to extract metals associated with individual fractions—that is, (step 1) the acid-soluble fraction, associated with exchangeable metals and bound with carbonate; (step 2) the reducible fraction, associated with metals bound to iron and manganese oxides; (step 3) the oxidizable fraction, including metals bound to organic matter and sulfides. After each extraction stage, the obtained residue was rinsed with deionized water, centrifuged, and subjected to another stage of extraction. In order to control the quality of the obtained results, an additional step was introduced to this procedure, which consisted of the digesting of ash with 10 mL of 65% $HNO_3$ and 2 mL of $H_2O_2$. The concentrations of metals in all the extracts were determined via ICP-MS.

**Table 1.** BCR speciation protocol.

| Extraction Step | Fraction | Extractant |
|---|---|---|
| I | Acid soluble: exchangeable metals bound with carbonates. | 0.11 M $CH_3COOH$<br>S/L = 1:40<br>16 h shaking 30 rotation/min |
| II | Reducible: metals bound to Fe and Mn oxyhydroxides. | 0.1 M $NH_2OH \cdot HCl$, pH 2 ($HNO_3$)<br>S/L = 1:40<br>16 h shaking 30 rotation/min |
| III | Oxidisable: metals bound to organic matter and sulfides. | 30% $H_2O_2$ per 1 h, then<br>1M $CH_3COONH_4$, pH 2 ($HNO_3$),<br>S/L = 1:50<br>16 h shaking 30 rotation/min |
| IV * | Residual: lithogenous, non-silicate bound metals. | 65% $HNO_3$ + $H_2O_2$ |

* Additional and recommended step.

The quality of the obtained results of all analyses was ensured by performing them according to a standard certified analytical quality control procedure according to PN-EN ISO 17294-1:2007 [51]. In order to further ensure the quality of the results obtained, reagent blanks and certified reference material (fly ash from pulverized coal, BCR 038) were used (including BCR protocol). The analytical bias was found to be statistically insignificant ($p$ = 0.05). The uncertainty of the obtained results is provided in Tables 2 and 3.

**Table 2.** Chemical composition of fly ash from biomass combustion in fluidized bed boilers.

| Parameter | | Fly Ash from CFB 1 | Fly Ash from CFB 2 | Fly Ash from BFB 1 | Fly Ash from BFB 2 | Threshold Values for Mineral Fertilizers (mg/kg) * | Minimum Nutrient Content for Mineral Fertilizers (wt%) * |
|---|---|---|---|---|---|---|---|
| V | | 22.6 ± 8.9 | 24.8 ± 9.7 | 14.3 ± 5.6 | 18.9 ± 7.4 | - | - |
| Cr | | 50.0 ± 20.8 | 44.5 ± 18.5 | 48.5 ± 20.1 | 53.6 ± 22.2 | - | - |
| Mn | | 2315 ± 801 | 2299 ± 795 | 5698 ± 1972 | 7157 ± 2476 | - | - |
| Co | | 6.17 ± 2.42 | 5.31 ± 2.09 | 4.31 ± 1.69 | 4.72 ± 1.85 | - | - |
| Ni | | 33.8 ± 10.9 | 27.2 ± 8.8 | 17.3 ± 5.6 | 21.0 ± 6.8 | - | - |
| Cu | | 112 ± 38 | 92.9 ± 31,7 | 146 ± 50 | 86.8 ± 29.6 | - | - |
| Zn | (mg/kg) | 325 ± 131 | 337 ± 135 | 583 ± 234 | 593 ± 238 | - | - |
| As | | 15.9 ± 4.9 | 6.41 ± 1.99 | 6.82 ± 2.12 | 7.85 ± 2.44 | 50 | - |
| Cd | | 6.12 ± 2.01 | 6.10 ± 2.00 | 8.14 ± 2.67 | 8.15 ± 2.67 | 50 */8 **/5 *** | - |
| Sn | | 7.50 ± 2.85 | 3.63 ± 1.38 | 1.02 ± 0.39 | b.d.l. | - | - |
| Sb | | 2.25 ± 0.68 | 2.67 ± 0.81 | 0.050 ± 0.015 | 0.800 ± 0.242 | - | - |
| Tl | | 1.15 ± 0.42 | 0.945 ± 0.347 | 2.11 ± 0.77 | 2.90 ± 1.06 | - | - |
| Pb | | 129 ± 45 | 71.3 ± 24.7 | 61.7 ± 21.4 | 51.4 ± 17.8 | 140 */200 **/600 *** | - |
| Hg | | 0.086 ± 0.023 | 0.064 ± 0.017 | 0.220 ± 0.059 | 0.240 ± 0.064 | 2 | - |
| $P_2O_5$ | | 2.00 ± 0.40 | 2.38 ± 0.48 | 3.57 ± 0.74 | 4.63 ± 0.93 | - | 2 |
| P | | 0.880 | 1.04 | 1.57 | 2.04 | - | - |
| $K_2O$ | | 6.20 ± 0.37 | 6.88 ± 0.41 | 6.62 ± 0.40 | 8.24 ± 0.49 | - | 2 |
| K | | 5.14 | 5.71 | 5.49 | 6.84 | - | - |
| CaO | | 12.9 ± 2.3 | 14.1 ± 2.5 | 26.5 ± 4.8 | 24.8 ± 4.5 | - | - |
| Ca | (% mass) | 9.19 | 10.06 | 18.94 | 17.73 | - | - |
| MgO | | 3.77 ± 0.45 | 4.06 ± 0.49 | 2.97 ± 0.36 | 3.31 ± 0.40 | - | - |
| Mg | | 2.27 | 2.45 | 1.79 | 1.99 | - | - |
| $SO_3$ | | 4.59 ± 0.92 | 3.97 ± 0.79 | 4.16 ± 0.83 | 4.53 ± 0.91 | - | - |
| S | | 1.84 ± 0.18 | 1.59 ± 0.16 | 1.66 ± 0.16 | 1.81 ± 0.18 | - | 2 |
| N | | 0.040 ± 0.004 | 0.030 ± 0.003 | 0.020 ± 0.002 | 0.020 ± 0.002 | - | - |
| Cl | | 1.54 ± 0.16 | 1.16 ± 0.12 | 1.42 ± 0.15 | 1.32 ± 0.14 | - | - |
| pH | | | | | | - | - |
| PEW | (mS/m) | 12.07 | 10.5 | 20.4 | 18.77 | - | - |

* Max. concentration of contaminants in mineral fertilizers according to Dz.U.119.765 [59]; ** max. concentration of contaminants in fertilizing lime (liming agent) 8 mg of Cd per 1 kg of CaO, 200 mg of Pb per 1 kg of CaO; *** max. concentration of contaminants in fertilizing lime containing magnesium 5 mg Cd per 1 kg of CaO + MgO, 600 mg Pb 1 kg of CaO + MgO. b.d.l.—below detection limit of the analytical device (for the V, Cr, Mn, Co, Ni, Cu, Zn, As, Cd, Sn, Sb, Tl, Pb detection limit for ICP-MS ranges between 0.25 and 2500 mg/kg).

**Table 3.** Leachability of the elements from biomass ash.

| Leachable Concentration of Elements | | Fly Ash from CFB 1 | Fly Ash from CFB 2 | Fly Ash from BFB 1 | Fly Ash from BFB 2 |
|---|---|---|---|---|---|
| V | | b.d.l. | b.d.l | $0.0035 \pm 0.0013$ | $0.0022 \pm 0.0008$ |
| Cr | | $5.5 \pm 1.46$ | $5.2 \pm 1.37$ | $4.46 \pm 1.19$ | $4.99 \pm 1.33$ |
| Mn | | $0.01 \pm 0.002$ | $0.032 \pm 0.0064$ | $0.037 \pm 0.0079$ | $0.043 \pm 0.0092$ |
| Co | | b.d.l. | b.d.l. | $0.0022 \pm 0.00053$ | $0.0015 \pm$ |
| Ni | | $0.5 \pm 0.086$ | $0.4 \pm 0.068$ | $0.066 \pm 0.0114$ | $0.0136 \pm 0.0023$ |
| Cu | (mg/kg) | b.d.l. | b.d.l. | $0.385 \pm 0.073$ | $0.074 \pm 0.0141$ |
| Zn | | $0.21 \pm 0.076$ | $0.36 \pm 0.13$ | $0.59 \pm 0.135$ | $0.75 \pm 0.169$ |
| As | | $0.09 \pm 0.033$ | $0.05 \pm 0.018$ | $0.036 \pm 0.00131$ | $<0.001$ |
| Cd | | b.d.l. | b.d.l. | $<0.001$ | $0.00079 \pm 0.00021$ |
| Sn | | $0.12 \pm 0.0316$ | b.d.l. | $<0.001$ | $<0.001$ |
| Tl | | b.d.l. | b.d.l. | $0.0148 \pm 0.0049$ | $0.0361 \pm 0.0118$ |
| Pb | | $0.04 \pm 0.010$ | $0.0352 \pm 0.009$ | $0.602 \pm 0.155$ | $0.715 \pm 0.184$ |
| $Cl^-$ | | $13{,}168 \pm 1027$ | $10{,}439 \pm 814$ | $12{,}230 \pm 954$ | $14{,}850 \pm 1158$ |
| $SO_4^{2-}$ | | $18{,}200 \pm 3585$ | $27{,}720 \pm 5461$ | $23{,}320 \pm 4594$ | $27{,}170 \pm 5352$ |
| $PO_4^{3-}$ | | b.d.l. | b.d.l. | b.d.l. | b.d.l. |
| $NO_3^-$ | (mg/kg) | $147 \pm 11$ | $152 \pm 11$ | $140 \pm 10$ | $146 \pm 11$ |
| $Ca^{2+}$ | | $6209 \pm 1130$ | $4411 \pm 803$ | $6940 \pm 1263$ | $6135 \pm 1117$ |
| $Mg^{2+}$ | | $339 \pm 40$ | $88.8 \pm 10.4$ | $0.200 \pm 0.023$ | $0.200 \pm 0.023$ |
| $Na^+$ | | $97.4 \pm 11.3$ | $39.7 \pm 4.6$ | $87.6 \pm 10.16$ | $44.7 \pm 5.18$ |
| $K^+$ | | $38{,}842 \pm 2447$ | $23{,}036 \pm 1451$ | $29{,}864 \pm 1881$ | $32{,}989 \pm 2078$ |
| $K_{bioavailable}(K_2O)$ | (mg/100 g) | 4043 | 3520 | 3750 | 3000 |
| $P_{bioavailable}(P_2O_5)$ | | 2.5 | 1.8 | 2.1 | 2.2 |

b.d.l.—below detection limit.

The acute phytotoxicity test (Phytotoxkit, Tiger MicroBioTest) was employed in order to determine the possible or potential inhibition of the seed germination, IG [%], as well as the inhibition of the root elongation, IR, as a result of soil amended with fly ash. These types of plant germination tests are commonly used by other researchers [52–56] to determine the toxicity of certain substrates (such as fertilizers, sludge, compost, waste, or other soil amendments) on the root elongation of terrestrial plants after a specific time of exposure to a certain soil contaminant when compared to control soil. The methodology used in the study was in line with ISO Standard 11269-1: 2012 [57]. OECD soil (series no: OERS011217) was used as a control and reference sample. Seeds of monocotyledonous (*Avena sativa* L.) and dicotyledonous (*Lepidium sativum*) plants were selected for the test in accordance with the OECD/OCDE guidelines 208/2006, which state that it is necessary to conduct research on plants from various systematic units.

The ash additive to OECD soil was calculated as 2.5 tonnes of CaO amendment per hectare for a 0.25 m depth of soil. Ten seeds of indicator plants were sown both in the experimental trials and in the control sample, as described in the phytotoxic test method. Calculations were also made in accordance with the test instructions. All the tests were performed in 3 replications.

Seeds of a selection of plants were laid on a paper filter lying on the surface of moistened soil/soil with fly ash. The plates were enclosed and then placed vertically and incubated at 25 °C in a thermostatic cabinet in the dark for 5 days. After the incubation, digital photographs of the incubated plates were taken, the number of germinated seeds was counted, and the root length of the germinated plants was measured. Finally, the inhibition of seed germination *IG* [%], the inhibition of root elongation *IR* in soil [%], as well as the *GI* germination index were calculated according to the following equations:

$$\frac{G_A - G_B}{G_A} \times 100 = IG\ [\%]$$

where:

$G_A$—average number of seeds germinating on control soil (OECD);

$G_B$—average number of seeds germinating on experimental medium.

$$\frac{R_A - R_B}{R_A} \times 100 = IR\ [\%]$$

where:

$R_A$—average root length on control soil (OECD);

$R_B$—average root length on experimental medium.

$$\frac{G_B \times R_B}{G_A \times R_A} \times 100\% = GI$$

## 3. Results and Discussion

### 3.1. Physico-Chemical Properties of Fly Ash from Biomass Combustion in Fluidized Bed Boilers

The chemical compositions of fly ash from both installations as well as its physical parameters are presented in Table 2. All the examined fly ash was alkaline, and the pH varied between 10.7 and 13.07, however the samples from the BFB installation were more alkaline than those of the CFB plant. The electrical conductance was reported to be high in all samples, however fly ash from the bubbling fluidized bed was twice as conductive (18.77–20.4 mS/cm) as samples taken from the circulating fluidized bed (10.05–12.07 mS/cm). These results are consistent with the findings of other researchers reporting a strongly alkaline pH of biomass ash delivered from large-size installations. Dahl et al. [28] reports that the pH of biomass ash delivered from a 246 MW fluidized bed boiler was found to be in the 11.9 to 12.6 range; Uliasz-Bocheńczyk et al. [58] reported biomass ash to have a pH even more alkaline (pH 12.92). Żelazny and Jarosiński [13], in their research on evaluating the biomass ash from Połaniec (205 MW) energy plant as a possible fertilizer, also confirm that the pH of biomass ash was highly alkaline, at pH > 11, and they concluded that such conditions may promote a significant loss of ammonia and phosphorus from NPK biomass fertilizer, as a result of the decomposition of ammonia from the nitrate ammonia compound and the formation of phosphorus compounds insoluble in water. In regard to the conductivity, Wilczyńska-Michalik et al. [12], in their research on biomass fly ash from the same 205 MW energy plant as ours, reveal a comparable conductivity which is equal to 11.38 mS/cm.

### 3.2. Macro and Micronutrient Contents in Biomass Fly Ash in Accordance with Fertilizer Legislation

All the fly ash samples derived from both installations contained a substantial number of elements, macronutrients (P, K, S, Ca, and Mg), and micronutrients (Mn, Cu, Zn, Co) considered as being essential for plant growth. The concentration of nitrogen, however, in all the evaluated samples was negligible, and it varied from 0.02% to 0.04%. These results correspond with the outcomes of other authors reporting a low level of nitrogen in biomass fly ash obtained from a variety of feedstock and combustion technologies [59–61]. It can therefore be concluded that none of the examined biomass fly ash from fluidized bed boilers met the minimum 2% nitrogen content threshold required for mineral fertilizers, in accordance with Polish legislation [59]. The concentrations of the remaining macronutrients, such as K and P, were satisfactory. However, when considering EU legislation regarding fertilizers [38], it was found that here the content of K and P were to meet the minimum requirements for K fertilizers (i.e., min. 10% of soluble $K_2O$), for PK fertilizers (min. 18% of $P_2O_5 + K_2O$), as well as for NPK fertilizers (min 20% of $N + P_2O_5 + K_2O$). The results are then in agreement with the findings of Żelazny and Jarosiński [13], who conclude that that the sole use of this type of waste as a full-value fertilizer is not possible; however, this waste could be considered as a source of potassium for the purpose of a more complex type of fertilizer (NK, PK, or NPK) production.

The fly ash from our BFB power plant was richer in potassium, phosphorus, and CaO than the ash delivered from our CFB power plant. The BFB fly ash contained two times more phosphorus (mean con. of 1.8% P) than ash from the CFB plant (mean con. of 0.96% P). It also contained two times more CaO (on average, 25.65% CaO compared to 13.46% in CFB) and had a higher level of potassium (7.43% $K_2O$ in BFB fly ash, compared to 6.54% $K_2O$ in CFB). The MgO and sulfur content remained comparable for both the BFB and CFB fly ash. The higher content of P and K in fly ash from the BFB could be the result of incinerating a higher share of "agro" biomass (21% agro addition) in the BFB boiler compared to the CFB installation, which incinerated only a 20% mix of "agro" bio. Moreover, the "agro" biomass used by the BFB boiler consisted of a mixture of sunflower husks (18%) and an additional 3% straw pellets. Straw, according to the results of various authors [62,63], appears to be richer in K and P content than sunflower husks alone.

Comparing our results with the findings of Wilczyńska-Michalik et al. [12], it can be safely concluded that the macronutrient content is mostly convergent, especially with regard to $P_2O_5$ and $K_2O$. However, the concentration of CaO as well as S in the CFB fly ash differed more significantly compared to both findings. For the fly ash samples CFB1 and CFB2, the concentration of CaO ranged from 12.86 to 14.07 wt%, which was found to be lower than the CaO concentration reported by Wilczyńska-Michalik et al. [12] (18.56 wt% CaO). On the other hand, the fly ash examined during our study was richer in sulfur content (1.58–1.83 wt%) than the fly ash examined by Wilczyńska-Michalik et al. [12] (1.12 wt%). The differences in the macronutrient content between both studies can probably be attributed to a different biomass feedstock mixture being incinerated in the same power plant. However, further comparison is difficult due to the lack of detailed information regarding the type of biomass used and its composition (especially undefined by the authors is the 20% agricultural feedstock addition).

The results of the macronutrient content of fly ash from BFB boilers were then referenced with the results of other researchers studying biomass ash obtained under comparable technological conditions. For example, Nurmesniemi et al. [20], who studied fly ash originating from large BFB power plant (115 MW) boilers, when incinerating clean forest biomass with an addition of 3% wastewater sludge reported comparable contents for Ca and P but a lower content of Mg 1.79–1.99% and a lower sum of P + K compared to the results obtained from our study performed on samples taken from a BFB installation. Our findings on the macronutrient composition of BFB ash revealed substantially more K (5.49–6.84 % mass) compared to the 3.9% reported by Nurmesniemi et al. [20], and even three times more CaO, $K_2O$, and $P_2O_5$ compared to the results reported by Wilczyńska-Michalik et al. [12], which they obtained by studying a much smaller (76.5 MWt) BFB installation and incinerating 100% agricultural residues.

A broader comparison of our results with the findings of other researchers [6–8,12,15,60,61] studying a multitude of fly ash from a variety of biomass feedstock and installations confirms that biomass incineration by-products are a quite heterogenous type of waste. For instance, in the very broad research conducted by Zając [7], it was found that the nutritional composition of fly ash derived from burning wood biomass, energy crops, agricultural biomass, and forest and agri-food industry waste varies significantly depending on the feedstock used (e.g., the P content (wt%) varied from 0.26 to 3.2, the K (wt%) varied from 1.9 to 18.7, and the Ca (wt%) varied from 3.6 to 35).

In our research, a high content of chlorine was found in both types of ash (1.16–1.54 wt%), which, according to Jaworek et al. [64], is a characteristic "trade mark" of bio-ash compared to coal fly ash.

The total concentrations of macronutrients in the fly ash of this study decreased in the descending order of nutritional elements Ca > K > Mg > S > P > N. Detailed results of this composition are depicted in Table 2. The concentrations of individual macronutrients ranged from 0.88% to 2.04% for P, 5.14–6.84% for K, and 0.02–0.02% for N, as well as 1.58–1.83% for S, called "the fourth macroelement". The concentration of micronutrients, on the other hand (Mn, Cu, Zn, Co), varied more significantly within different types of fly ash, especially with regard to the manganese content, ranging from 2299 up to 7157 mg/kg. The concentration of this essential nutrient (Mn) was three times higher in fly ash from the BFB than that of the fly ash from the CFB installation. Similarly, the concentration of Zn in the fly ash from the BFB was found to be twice as high as that of the CFB fly ash, on average 588 mg/kg Zn and 331 mg/kg, respectively. The concentrations of the remaining micronutrients, such as Co and Cu, in ash was comparable and varied insignificantly between the two installations. After comparing the obtained micronutrient content results with the findings of Wilczyńska-Michalik et al. [12] for the same installation (205 MW), it was determined that our outcomes are in line with one another.

### 3.3. Non-Essential Elements and Contaminants

Besides nutritional elements, biomass fly ash waste also contains metals (V, Cr, Ni) that are non-essential for plants but at the same time could be considered beneficial for their growth when introduced in small amounts. For instance, Vanadium is not an essential element for plants, however it can stimulate growth and chlorophyll formation when added in small quantities [65]. At the same time, this metal can also be toxic to plants when present in elevated concentrations. Some authors [66] report that the addition of Vanadium to flovo-aquic soil in amounts exceeding 30 mg/kg significantly decreases the yields of shoots and roots. The concentration of Vanadium did not exceed 24.8 mg/kg in all the examined fly ash obtained from both the CFB and BFB installations. However, since those values for V concentrations are not so far apart, further research on that issue should be conducted.

Chromium is also a non-essential element which is potentially detrimental to plants, causing oxidation stress and initiating the degradation of photosynthetic pigments, consequently resulting in a decline in plant growth. Although conversely, as indicated by Shanker [67], Cr can actually enhance the growth of certain plant species at lower concentrations. The concentration of Cr in the fly ash of our study was not elevated and was comparable between the two installations, and it was not elevated, varying between 44.5 and 53.6 mg/kg. The content of this element was thus much lower than, e.g., that obtained by Schiemenz and Eichler-Löbermann [15] for rape meal ash derived from laboratory fluidized bed combustion. Although the concentrations of Cr and Ni are not specified by fertilizer legislation, it is noteworthy to point out that, in all of the studied biomass ash samples, the concentrations of these heavy metals were not elevated and were found to be within the upper threshold values established for the 1st quality soil group in accordance with Polish legislation [68]. Similarly, the concentration of Ni was rather low, and thus did not exceed the threshold values established for type 1 classification. It ranged from 17.3 mg/kg in the fly ash from BFB up to 33.8 mg/kg for the CFB fly ash. Therefore, fly ash from both types of fluidized bed boilers should be regarded as not potentially harmful to plants' growth with regard to V, Cr, and Ni contamination.

Biomass fly ash also contains highly phytotoxic elements which do not play any role in plant metabolism and are simply considered as contaminants (As, Cd, Pb, Sb, Tl, and Hg). These metals are not biologically essential for plants, and they are highly phytotoxic at certain threshold values. The maximum permissible levels of As, Cd, Pb, and Hg in mineral fertilizers are regulated by appropriate legislation [59]. All of the examined fly ash samples did not exceed the threshold values established by the above-mentioned regulation regarding the content of contaminants. The concentration of as in the samples ranged from 6.41 to 15.9 mg/kg, thus not exceeding the 50 mg/kg limit. Furthermore, the fly ash samples contained a low level of Cd, ranging from 6.1 to 8.15 mg/kg; Pb, ranging from 51.4 to 129 mg/kg; Sb, ranging from 0.05 to 2.67 mg/kg; and Tl, ranging from 0.94 to 2.9 mg/kg, as well as a low concentration of Hg, ranging from 0.086 mg/kg in the CFB ash to 0.24 mg/kg in the ash delivered from the BFB installation. However, after comparing the above concentrations with the threshold values established for liming agents (fertilizing lime and fertilizing lime containing magnesium), it can be concluded that only fly ash from the circulating fluidized bed installation can legally be used as a direct soil liming amendment, whereas the content of Cd (8.14–8.15 mg/kg) in the ash from the bubbling bed exceeds both the maximum permissible concentrations of 8 mg of Cd per 1 kg of CaO as well as 5 mg of Cd per 1 kg of CaO + MgO.

On the other hand, when considering the cumulative concentration of elements in ash regarded as contaminants (As, Cd, Pb, Hg, Tl, and Sb), as well as toxic and non-essential elements (such as V, Cr, and Ni), it was found that the fly ash derived from CFB installations is more contaminated with metals than the fly ash derived from the BFB installation. The cumulative concentration of metals in the CFB samples reached an average of 227.9 mg/kg, and an average of 162.2 mg/kg in the BFB samples.

A comparison of our results with those presented by Wilczyńska-Michalik et al. [12] for the same installation revealed that, while the contents of Co, Ni, and Cu in the CFB fly ash from both studies were comparable, the results of the concentrations of Cr, Cd, Tl, and Pb varied greatly, by as much as three-fold, depending on the metal. For example, the fly ash obtained by Wilczyńska-Michalik et al. [12] was much more contaminated with Cr and Cd. Furthermore, they reported a Tl concentration 10 times lower and a Pb content three times lower than the ones observed in this study. This was the case despite all the samples originating from the very same circulated bed boiler installation. However, as previously stated the above-mentioned variations in our study compared with that of Wilczyńska-Michalik et al. [12] may result from using a different mixture of biomass feedstock for the incineration process in the CFB power plant, or may even be attributed to using different metal digestion protocols. It has to be noted that different ash mineralization protocols, employing various extraction liquids and various equipment, can greatly influence the recovery rates of metals from fly ash.

The BFB ash results were mostly consistent with the findings of other researchers, such as Dahl et al. [28] and Nurmesniemi et al. [20,31], who conducted similar studies on the content of metals in fly ash derived from large-size BFB installations in Finland (296, 246, and 115 MW). Comparing their outcomes with the results of this study on BFB ash from a 183 MW installation, some conclusions may be drawn. The first is that the BFB1 as well as BFB2 samples of fly ash contained substantially less contaminants such as Cr and As. Second, more Cd and Zn was, however, determined in our BFB ash compared to the fly ash originating from large power plants in Finland (115, 246, and 296 MW) [20,28,31].

### 3.4. Bioavailability of Elements from Fly Ash

The total concentration of metals in fly ash does not deliver sufficient information on the real mobility and bioavailability of these elements. The bioavailability of metals in the soil–plant environment is a very complex issue governed by multiple factors, such as the pH; redox potential; organic content of the substrate; total content of metals; speciation; concentration of organic and inorganic ligands, including humic and fulvic acids; soil texture; clay content; microbial activity; or simply the coexistence of synergetic or antagonistic metals. In such a complex substrate as soil, most of all the above-mentioned factors are interrelated and can vary in wide ranges. Metals considered as both macro and micronutrients, unlike organic matter, are not metabolically degradable, and by changing their chemical forms from soluble to insoluble (due to the above-mentioned factors), they can stay in the ecosystem for tens or even hundreds of years [69–72].

Considering the fact that metals in the fly ash are not permanently fixed, an extended study on the bioavailability of elements using aqueous leaching tests as a well speciation study, based on three-step sequential extraction proposed by the European Community Bureau of Reference (BCR), was deemed necessary. A one-stage aqueous leaching test was chosen because it is most commonly used to pre-characterize the toxic effect of the substrate and to deliver preliminary information on easily soluble forms of metals in ash. Sequential extraction protocols were further used to broaden the scope of the research by providing information on the main phases of metals in which the metals are bound in ash, thus delivering results on its potential anthropogenic and lithogenic origin. The potential mobility of the trace elements in all the examined fly ash is summarized in Table 3.

The aqueous leaching test revealed the enhanced leachability of sulfate ions for all fly ash samples, ranging from 18,200 up to 27,720 mg/kg. Such a high mobility of sulfates is a consequence of the ash composition, which also contains a substantial amount of waste gypsum, a by-product of the desulfurization process which is incorporated in the waste stream during fluidized bed combustion. Elemental sulfur is absorbed by plants when oxidized to sulfate ions. This element in its bioavailable form is highly favorable in all fertilizers, since sulfur is essential for plant growth and functioning [73,74], it provides proper nutrition for plants, resulting in increased yields, and improving their quality [75]; is responsible for the resistance of plants to biotic and abiotic stresses; and governs and controls proper nitrogen metabolism. Thus, when properly supplemented, it allows a reduction in the applied doses of nitrogen fertilizers. A very high content of chloride anions ranging from 10,440 to 14,850 mg/kg was also found in all the fly ash leachates from both installations. Chloride is an essential micronutrient to plants, but only in small amounts. However, when it appears in such extreme concentrations of chloride anions, as found in both types of fly ash, it can be potentially detrimental to plants, causing salinity stress and thus a reduction in yield, water uptake, or photosynthetic capacity due to chlorophyll degradation [75]. At the same time, it should be stressed that chloride and sulfate anions are antagonistic, so the excess of one anion can cause the limited availability of another anion in fly ash. Leaching tests results also revealed the high mobility and bioavailability of other macronutrients, such as potassium (23,030 to 38,840 mg/kg) and calcium (4410 to 6940 mg/kg).

Vassiliev et al. [5] reports that the high leachability of Cl, S, or Ca from biomass ash may result from the content of highly soluble chlorides, such as sylvite; halite; sulfates (e.g., ettringite, gypsum, anhydrite, etc.); or carbonates, such as calcite, dolomite, etc.

A relatively low mobility of magnesium was reported, varying from 2 to 330 mg/kg, as well as a negligible amount of easily soluble phosphates when compared to their total content in the ash. It has to be noted that phosphorus is available for plants in various forms, including active phosphorus (present in the soil solution in the form of phosphoric acid dissociation ions), mobile phosphorus (i.e., its compounds are soluble in weak acids), and "spare" phosphorus (in the form of various types of apatites). Total phosphorus, then, is the sum of all its above-mentioned forms. The leaching test results revealed a lack of active and easy soluble phosphorus in ash, undetectable using the IC method. More accurate research on other mobile and bioavailable forms of this element in fly ash (including both organic and inorganic forms) was conducted for this reason. The concentration of bioavailable phosphorus determined using the Egner–Riehm method was rather low in all the samples and ranged from 2.0 to 2.2 mg/100 g of $P_2O_5$ (20–22 mg/kg). Our results are therefore consistent with the findings of other authors [15,20,76,77] reporting the poor water solubility and bioavailability of phosphorus from biomass ash to plants. However, as indicated by Schiemenz and Eichler-Löbermann [15], the bioavailability of phosphorus is governed mainly by the soil pH, so the better solubility of calcium phosphates under acidic pH conditions can enhance the effect of biomass ash. The low water solubility of phosphorus should also be regarded as a positive outcome, since it limits the risk of the uncontrolled leaching of this element from the ash to the soil when considered as a soil amendment for forestry use. The remaining trace metals—V, Mn, Co, Zn, Tl, Pb, and As—in both types of fly ash were also poorly soluble in water, and their concentrations in leachates did not exceed 1% when compared to the total content of metals in ash. The only exception was Cr, which was leached out from both types of ash (CFB and BFB) in amounts accountable for about 10% of its total content in each fly ash sample. Detailed results from the aqueous leaching test are provided in Table 3.

Speciation of Metals in Fly Ash

A BCR speciation study was conducted with the use of more aggressive sets of reagents in order to reveal information about Cd, Zn, and Pb binding forms in certain types of fly ash and their probable origin. Detailed results are depicted in Figures 1–3.

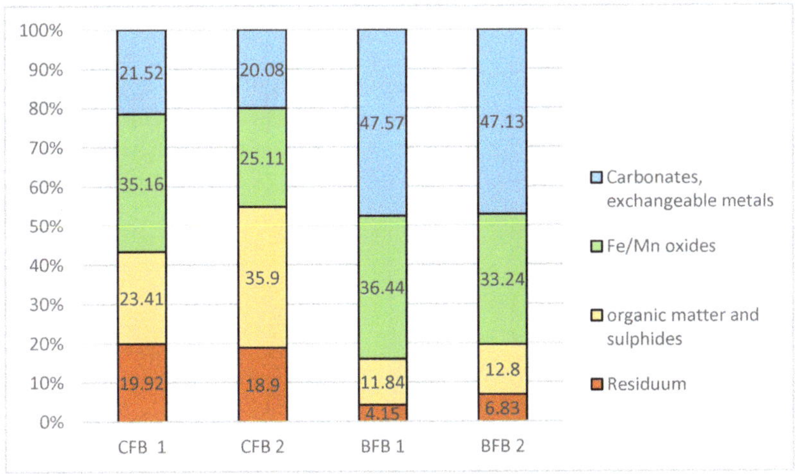


**Figure 1.** Speciation of cadmium in fly ash.

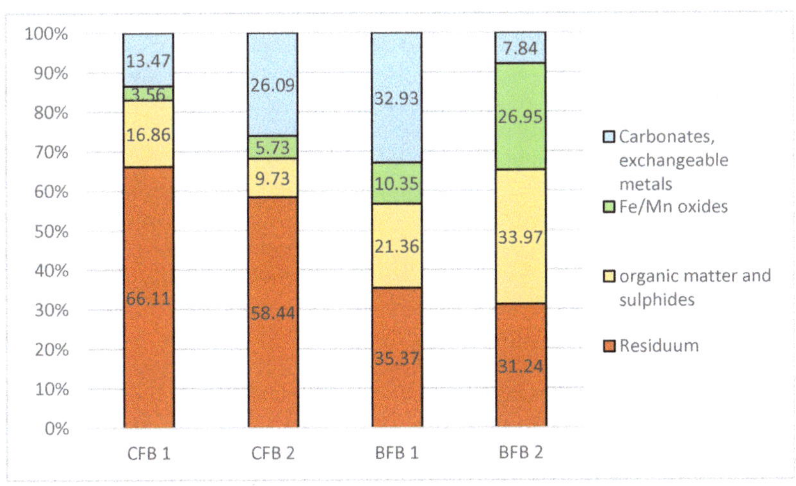


**Figure 2.** Speciation of Pb in fly ash.

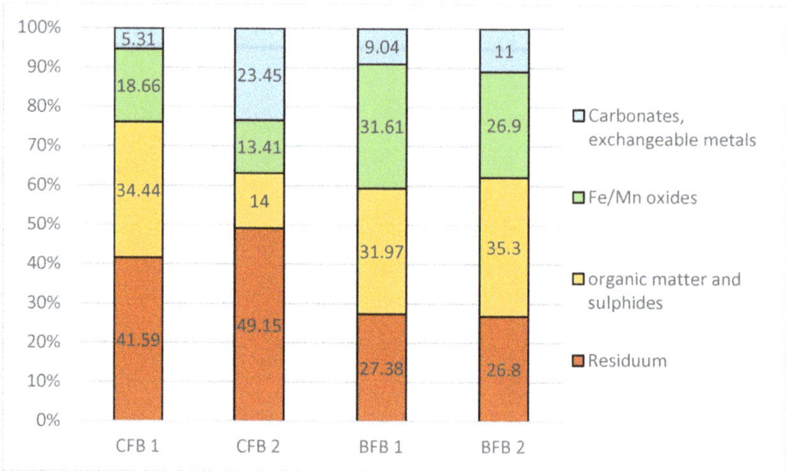



**Figure 3.** Speciation of Zn in fly ash.

The cadmium contents in individual fractions of both CFB and BFB fly ash differ within samples. In the fly ash from the CFB, cadmium was bound with all three fractions almost in equal amounts; however, in ash CFB 1 this metal was associated mostly with reducible fractions of amorphous Fe/Mn oxides (35%), whereas in the second sample of ash, CFB2, delivered from the same installation, Cadmium was predominantly bound with the oxidizable fraction, not susceptible to leaching and associated with organic matter and sulfites, as well as with the mineral residuum phase (total of 54.8%). In both samples of fly ash from the CFB installation, about 20–21% of the cadmium was bound with the labile and acid soluble phase associated with carbonates, from which metals can easily by remobilized under, e.g., the dropping of pH. In the fly ash collected from the bubbling bed installation, cadmium was much more mobile and potentially bioavailable, since the prevailing amount of this element was bound with easily exchangeable fractions (about 47%), as well as with the reducible phase (about 33% to 36% of Cd). As indicated by the authors [78–80], the reducible fraction acts as a sink to contaminants because Fe/Mn oxides are present as coatings on mineral surfaces or clay particles in the soil matrix, and consequently contaminants can be remobilized from that phase under redox conditions. Consequently, only about of 4% to 6% of the cadmium was then immobile and fixed with the residuum, and about 12% of the Cd was associated with the oxidizable phases of organic matter and sulfides.

Conducting a fractionation study revealed that the prevailing amount of lead (58–66%) in both samples of CFB fly ash is not bioavailable, since it is bound in the residual fraction, while 10% to 17% of this element is also associated with organic matter and sulfides. This indicates that lead still can be potentially released under oxidizing conditions. To the smallest extent (3.56–5.73%), the lead in CFB ash is bound with the reducible fraction associated with Fe/Mn oxides, and it can be susceptible to release under reducible conditions. The remaining amount of lead, approximately 13% to 26%, is weakly absorbed by carbonates and can be easily released by ion-exchangeable processes, for example.

In fly ash from the bubbling bed boiler, lead is much more liable to leach when considering the first two phases combined—that is, an acid-soluble exchangeable phase, as well as a reducible phase associated with Fe/Mn amorphous oxides. In the first sample, BFB 1, almost 33% of lead is easily releasable, and an additional 10.35% can be leached out when the soil conditions change from oxic to anoxic, while in sample BFB2 the proportions are the opposite and lead is bound in 27% of the reducible phase and only approximately 8% is easily releasable when the pH of the soil or other medium drops. Only 31% to 35% of lead is safely fixed within the mineral residuum fraction, and the remaining 22% to 34% is bound to organic matter and sulfides (as depicted in Figure 2).

A total of 26.8% to 49% of Zn in both types of fly ash should be regarded as not mobile and not available to plants, since it is safely bound with the mineralogical fraction of the ash residuum (Figure 3). However, the remaining amount of Zn, being an essential micronutrient, is potentially available under either acid-soluble, reducible, or oxidizable conditions. In both the CFB fly ash, zinc is predominantly bound with the residual phase. The content of this element in the carbonates and exchangeable fraction as well as the oxidizable fraction differs greatly within individual samples, and it ranges from 5.31% Zn in the exchangeable fraction in CFB1 to 23.45% in the second sample of the CFB ash, as well as 34.44% Zn in the oxidizable fraction of CFB1 when compared to only 14% in the CFB2 ash. In the reducible fraction associated with Fe/Mn oxides, the content of Zn is, however, comparable, and it ranges from 13.43% to 18.66%. In both the BFB ash samples, the concentrations of Zn in certain fractions are comparable. A total of 26–27% of Zn is bound with the residuum, 32–35% is associated with organic matter and sulfites, 27–31.6% of Zn is associated with the Fe/Mn oxyhydroxides, and only 9–10% of Zn is easily soluble and bioavailable from both BFB ash.

## 3.5. Acute Toxicity of Fly Ash Amendments to Plants Germination and Growth

The results of the acute toxicity test are summarized in Tables 4 and 5 and are depicted in Figure 4.

Conducting research on the potential toxic influence of biomass fly ash amendment on plants revealed no inhibition of the seed germination of *Lepidium sativum* in soil with the addition of both CFB fly ash and BFB2, whereas the inhibition of *Avena sativa* seeds from 3.3% to 6.7% was found in all mixtures of control OECD soil and the addition of fly ash, except for the CFB2 addition, where 10 out of 10 seeds of both plants germinated.

**Table 4.** Percentage inhibition of the seed germination, IG, calculated based on the average number of germinated seeds.

| Mixtures | An Average Number of Germinated Seeds | | The Percentage Inhibition of Seed Germination IG (%) | |
|---|---|---|---|---|
| | *Avena sativa* | *Lepidium sativum* | *Avena sativa* | *Lepidium sativum* |
| OECD control soil | 10.00 | 10.00 | - | - |
| OECD + BFB1 | 9.67 | 9.67 | 3.30 | 3.30 |
| OECD + BFB2 | 9.67 | 10.00 | 3.30 | 0.00 |
| OECD + CFB1 | 9.33 | 10.00 | 6.70 | 0.00 |
| OECD + CFB2 | 10.00 | 10.00 | 0.00 | 0.00 |

**Table 5.** An average root length of the germinated seeds.

| Mixtures | An Average Root Length of Germinated Seeds (mm) | | Root Growth Inhibition (%) | |
|---|---|---|---|---|
| | *Avena sativa* | *Lepidium sativum* | *Avena sativa* | *Lepidium sativum* |
| OECD control soil | 95.0 | 55.0 | - | - |
| OECD + BFB1 | 102.7 | 56.3 | −8.10 * (8.10% stimulation) | −2.36 (2.36% stimulation |
| OECD + BFB2 | 93.3 | 50.0 | 1.79 | 9.09 |
| OECD + CFB1 | 94.1 | 63.0 | 0.95 | −14.55 (14.55% stimulation) |
| OECD + CFB2 | 87.5 | 57.8 | 7.89 | −5.09 (5.09% stimulation) |

* Negative inhibition stands for stimulation (according to IO ISO Standard 11269-1: 2012 [57].

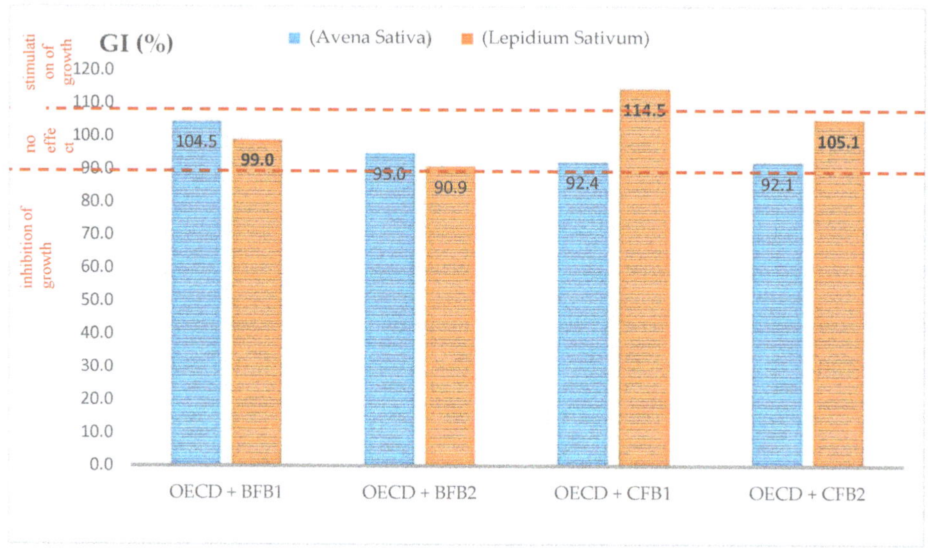


**Figure 4.** Germination indices for *Avena sativa* and *Lepidium sativum*.

The stimulation of *Lepidium sativum* root growth was found as a result of fertilizing the soil with all ash amendments except for BFB2, whose addition caused 9% root inhibition. The best growth stimulating agent was ash from the circulating bed boiler (CRB1), which caused a 14.55% stimulation of root growth. When analyzing the *Avena sativa* root growth elongation, the results were exactly the opposite, and stimulation was found only in one sample (amended with BFB1 ash), whereas in the remaining samples the inhibition of root growth appeared in 0.95% to 7.89%. It can therefore be safely concluded that the amendment of soil with ash from the bubbling bed boiler installation was the most suitable fertilizing amendment, because it did not cause any inhibition of growth for both plants and resulted in a slight stimulation of root growth elongation (2.36% to 8.1%), depending on the plant species.

Furthermore, the Germination Index (GI), which is considered to be the most important parameter indicating the possible toxic effect of any substrate on plant growth, was calculated based on the number of germinated seeds and the root length of germinated seeds. The results depicted in Figure 4 clearly show that none of the biomass fly ash additive had any negative effect on the germination and growth of *Avena sativa* as well as *Lepidium sativum*. For both plants, the calculated GI germination rate ranged from 90.9% to 114.5%. For one sample containing an addition of fly ash from the CFB installation, the germination index indicated even a stimulation of the *Lepidium sativum* growth.

## 4. Conclusions

Based on the results of this study, it can be concluded that the biomass fly ash obtained from the bubbling fluidized bed boilers was richer in potassium, phosphorus, carbonates, and micronutrients than the ash delivered from the circulating fluidized bed boilers. The BFB ash also contained cumulatively less contaminants such as V, Cr, Ni, As, Cd, Sb, Tl, Pb, and Hg than that from the CFB. However, when comparing the results of the Cd content with the threshold values established for liming agents (fertilizing lime and fertilizing lime containing magnesium), it becomes evident that only the fly ash from circulating fluidized bed (CFB) installation can legally be used as a direct soil liming amendment, since the Cd content of ash from the bubbling bed boilers (8.14–8.15) exceeds both the maximum acceptable concentrations of 8 mg of Cd per 1 kg of CaO as well as 5 mg of Cd per 1 kg of CaO + MgO. The difference in concentration between both types of fly ash can be attributed to the different feedstock mixture ratio used in both plants. In the BFB boilers, a higher ratio of "agro" biomass was incinerated on top of the 79% forest biomass, including 3% straw pellets and 17% sunflower husks, whereas, in the case of CFB, 80% forest biomass as well as 20% sunflower husks were incinerated. Moreover, the BCR speciation study revealed that fly ash from both installations met the threshold values and minimum requirements set for mineral fertilizers, with the exception of the nitrogen content, which was found to be below 2 wt% in each sample. The phosphorus in all fly ash samples was, however, very poorly extractable and not easily bioavailable to plants. Active phosphorus in the form of phosphate anions was leached in negligible amounts from both the BFB as well as CFB ash, and the content of bioavailable phosphorus $P_2O_5$ (soluble in weak acids) ranged from 20–22 mg/kg. On the other hand, the leaching test results revealed the very high mobility and bioavailability of other macronutrients, such as potassium (from 23,030 to 38,840 mg/kg), calcium (from 4410 to 6940 mg/kg), and sulfur in terms of sulfates (from 18,200 to 27,720 mg/kg). The contaminants in ash (Pb, Cd, Ce, V, Ni, As, Tl) were not easily mobile in the biomass ash, since their concentrations in aqueous leachates were negligible. A speciation BCR study revealed that almost 50% of Cd is highly mobile and bioavailable in BFB ash, since it is associated with exchangeable fractions and carbonates. Additionally, 33% to 36% of Cd is also potentially bioavailable under reducible conditions (bound with Fe/Mn oxides). Only 4–6.8% of Cd is safely bound with the residuum. In the CFB ash, approximately 50% of the Cd is available (bound with carbonates as well as Fe/Mn oxides), and can easily by remobilized under lower pH or redox conditions. Only approximately 20% of Cd is fixed in the residual fraction. The prevailing amounts of Pb (58–66%) in both samples of CFB fly ash are not bioavailable, since they are bound in the residual fraction. A total of 10% to 17% of this element is associated with organic matter and sulfides, thus indicating that lead can still be potentially released under oxidizing conditions. Zinc, a valuable micronutrient, is more bioavailable in the CFB ash than in the BFB biomass ash.

The results of an acute toxicity test also confirm that biomass fly ash amendment to soil does not have any toxic influence on plant germination and growth, despite a very high concentration of chloride anions, which are potentially detrimental to plants when appearing in such concentrations as found in both types of ash. Considering the favorable physico-chemical properties of biomass ash, especially pertaining to its high content of CaO, potassium, and other macronutrients, it is justifiable to further investigate the possible utilization approach of this particular waste (e.g., as an additive to fertilizers or as a soil improving agent), especially since this utilization approach will fit well into the waste management hierarchy as well as the circular economy policy currently promoted by EU countries. Moreover, upcoming new fertilizer regulations, which promote the idea of using organic, bio-waste, or recycled fertilizers such as biomass ash in top soils, serve to make it more justifiable to perform further research on this specific utilization approach. This is even more true when one additionally considers using this material for reclamation purposes or as a forest fertilizer. The current landfilling of biomass ash should be regarded as a highly unfavorable solution and quite simply wasteful. It, by extension, means that the entire effort put into producing energy from ecological sources was simply wasteful as well, particularly considering that this type of biomass ash, when properly treated, can easily return to the soil, thus closing the natural biogeochemical cycle.

**Author Contributions:** Conceptualization, E.J.-K.; analysis, E.J.-K. and J.P.; resources, J.P. and E.J.-K.; writing—original draft preparation, E.J.-K.; writing—review and editing, E.J.-K.; visualization (figure and table preparation), supervision, E.J.-K.; funding acquisition, E.J.-K.; data curation, E.J.-K. All authors have read and agreed to the published version of the manuscript.

**Funding:** The work was supported by the National Centre for Research and Development (NCBR) as research project no. PBS3/A2/21/2015. Publication fee was funded by the Polish National Agency for Academic Exchange under the International Academic Partnerships Programme from the project "Organization of the 9th International Scientific and Technical Conference entitled Environmental Engineering, Photogrammetry, Geoinformatics—Modern Technologies and Development Perspectives" 17–20 September 2019, Lublin, Poland.

**Conflicts of Interest:** The authors declare no conflict of interest. The funders had no role in the design of the study, in the collection, analysis, or interpretation of data, in the writing of the manuscript, or in the decision to publish the results.

## References


1. Dz.U. *2014 poz. 1923. Regulation of the Minister of the Environment of 9 December 2014 on the Waste Catalog;* Journal of Laws of the Republic of PolandJournal of Laws: Warsaw, Poland, 2014.
2. Pallarès, D.; Johnsson, F. Macroscopic modelling of fluid dynamics in large-scale circulating fluidized beds. *Prog. Energy Combust. Sci.* **2006**, *32*, 539–569. [CrossRef]
3. Leski, K.; Luty, P.; Łucki, A.; Jankowski, D. Application of circulating fluidized bed boilers in the fuel combustion process. *Tech. Trans.* **2018**, *4*, 83–96.
4. Wang, H.; Bolan, N.; Hedley, M.; Horne, D. Potential Uses of Fluidised Bed Boiler Ash (FBA) as a Liming Material, Soil Conditioner and Sulfur Fertilizer. In *Coal Combustion Byproducts and Environmental Issues;* Sajwan, K.S., Twardowska, I., Punshon, T., Alva, A.K., Eds.; Springer: New York, NY, USA, 2006. [CrossRef]
5. Vassilev, S.V.; Baxter, D.; Andersen, L.K.; Vassileva, C.G. An overview of the chemical composition of biomass. *Fuel* **2010**, *89*, 913–933. [CrossRef]
6. Vassilev, S.V.; Baxter, D.; Andersen, L.K.; Vassileva, C.G. An overview of the composition and application of biomass ash. Part 1. Phase–mineral and chemical composition and classification. *Fuel* **2013**, *105*, 40–76. [CrossRef]
7. Zając, G.; Szyszlak-Bargłowicz, J.; Gołębiowski, W.; Szczepanik, M. Chemical characteristics of biomass ashes. *Energies* **2018**, *11*, 2885. [CrossRef]
8. Ciesielczuk, T.; Rosik-Dulewska, C.; Kochanowska, K. The Influence of biomass ash on the migration of heavy metals in the flooded soil profile—Model experiment. *Arch. Environ. Prot.* **2014**, *40*, 3–15. [CrossRef]
9. Ciesielczuk, T.; Kusza, G.; Nemś, A. Nawożenie popiołami z termicznego przekształcania biomasy źródłem pierwiastków śladowych dla gleb. *Ochr. Środowiska Zasobów Nat.* **2011**, *49*, 219–227. (In Polish)
10. Uliasz-Bocheńczyk, A.; Pawluk, A.; Pyzalski, M. Charakterystyka popiołów ze spalania biomasy w kotłach fluidalnych. *Miner. Resour. Manag.* **2016**, *32*, 149–162. [CrossRef]
11. Uliasz-Bocheńczyk, A.; Mokrzycki, E. The elemental composition of biomass ashes as a preliminary assessment of the recovery potential. *Miner. Resour. Manag.* **2018**, *34*, 115–132. [CrossRef]
12. Wilczyńska-Michalik, W.; Gasek, R.; Michalik, M.; Dańko, J.; Plaskota, T. Mineralogy, chemical composition and leachability of ash from biomass combustion and biomass–coal co-combustion. *Mineralogia* **2018**, *49*, 67–97. [CrossRef]
13. Żelazny, S.E.; Jarosiński, A. The evaluation of fertilizer obtained from fly ash derived from biomass. *Miner. Resour. Manag.* **2019**, *35*, 139–152.
14. Olanders, B.; Steenari, B.-M. Characterization of ashes from wood and straw. *Biomass Bioenergy* **1994**, *8*, 105–115. [CrossRef]
15. Schiemenz, K.; Eichler-Löbermann, B. Biomass ashes and their phosphorus fertilizing effect on different crops. *Nutr. Cycl. Agroecosyst.* **2010**, *87*, 471–482. [CrossRef]
16. Patterson, S.J.; Acharya, S.; Thomas, J.; Bertschi, A.B.; Rothwell, R.L. Integrated soil and crop management: Barley biomass and grain yield and canola seed yield response to land application of wood ash. *Agron. J.* **2004**, *96*, 971–977. [CrossRef]
17. Mercl, F.; Tejnecký, V.; Száková, J.; Tlustos, P. Nutrient Dynamics in Soil Solution and Wheat Response after Biomass Ash Amendments. *Agron. J.* **2016**, *108*, 2222–2234. [CrossRef]
18. Nabeela, F.; Murad, W.; Khan, I.; Mian, I.A.; Rehman, H.; Adnan, M.; Azizullah, A. Effect of wood ash application on the morphological, physiological and biochemical parameters of *Brassica napus* L. *Plant Physiol. Biochem.* **2015**, *95*, 15–25. [CrossRef] [PubMed]
19. Ozolinčius, R.; Varnagirytě, I.; Armolaitis, K.; Karltun, E. Initial effects of wood ash fertilization on soil, needle and litterfall chemistry in a Scots Pine (*Pinus sylvestris* L). *stand. Balt. For.* **2005**, *11*, 21.
20. Nurmesniemi, H.; Manskinen, K.; Poykio, R.; Dahl, O. Forest fertilizer properties of the bottom ash and fly ash from a large-sized (115 MW) industrial power plant incinerating wood-based biomass residues. *J. Univ. Chem. Technol. Met.* **2012**, *47*, 43–52.
21. Ohenoja, K.; Pesonen, J.; Yliniemi, J.; Illikainen, M. Utilization of Fly Ashes from Fluidized Bed Combustion: A Review. *Sustainability* **2020**, *12*, 2988. [CrossRef]

22. Phongpan, S.; Mosier, A. Impact of organic residue management on nitrogen use efficiency in an annual rice cropping sequence of lowland Central Thailand. *Nutr. Cycl. Agroecosyst.* **2003**, *66*, 233–240. [CrossRef]

23. Ikpe, F.N.; Powell, J.M. Nutrient cycling practices and changes in soil properties in the crop-livestock farming systems of western Niger Republic of West Africa. *Nutr. Cycl. Agroecosyst.* **2002**, *62*, 37–45. [CrossRef]

24. Saletnik, B.; Bajcar, M.; Zaguła, G.; Czernick, M.; Puchalski, C. Influence of biochar and biomass ash applied as soil amendment on germination rate of Virginia mallow seeds (Sida hermaphrodita R.). Econtechmod. *An Int. Q. J.* **2016**, *5*, 71–76.

25. Owolabi, O.; Ojeniyi, S.; Amodu, A.; Hazzan, K. Response of cowpea, okra and tomato sawdust ash manure. *Moor J. Agric. Res.* **2003**, *4*, 178–182. [CrossRef]

26. Meller, E.; Bilenda, E. Wpływ popiołów ze spalania biomasy na właściwości fizykochemiczne gleb lekkich. *Polityka Energetyczna* **2012**, *15*, 287–292. (In Polish)

27. Ayeni, L.S.; Oso, O.P.; Ojeniyi, S.O. Effect of sawdust and wood ash application in improving soil chemical properties and growth of cocoa (*Theobroma cacao*) seedlings in the Nurseries. *Agric. J.* **2008**, *3*, 323–326.

28. Dahl, O.; Nurmesniemi, H.; Poykio, R.; Watkins, G. Heavy metal concentrations in bottom ash and fly ash fractions from a large-sized (246 MW) fluidized bed boiler with respect to their Finnish forest fertilizer limit values. *Fuel Process. Technol.* **2010**, *91*, 1634–1639. [CrossRef]

29. Ohenoja, K.; Tanskanen, P.; Wigren, V.; Kinnunen, P.; Körkkö, M.; Peltosaari, O.; Österbacka, J.; Illikainen, M. Self-hardening of fly ashes from a bubbling fluidized bed combustion of peat, forest industry residuals, and wastes. *Fuel* **2016**, *165*, 440–446. [CrossRef]

30. Dahl, O.; Nurmesniemi, H.; Pöykiö, R.; Watkins, G. Comparison of the characteristics of bottom ash and fly ash from a medium-size (32 MW) municipal district heating plant incinerating forest residues and peat in a fluidized-bed boiler. *Fuel Process. Technol.* **2009**, *90*, 871–878. [CrossRef]

31. Nurmesniemi, H.; Mäkelä, M.; Poykio, R.; Manskinen, K.; Dahl, O. Comparison of the forest fertilizer properties of ash fractions from two power plants of pulp and paper mills incinerating biomass-based fuels. *Fuel Process. Technol.* **2012**, *104*, 1–6. [CrossRef]

32. Pöykiö, R.; Nurmesniemi, H.; Keiski, R.L. Total and size fractionated concentrations of metals in combustion ash from forest residues and peat. *Proc. Est. Acad. Sci.* **2009**, *58*, 247–255. [CrossRef]

33. Pöykiö, R.; Nurmesniemi, H.; Perämäki, P.; Kuokkanen, T.; Välimäki, I. Leachability of metals in fly ash from a pulp and paper mill complex and environmental risk characterization for eco-efficient utilization of the fly ash as a fertilizer. *Chem. Speciat. Bioavailab.* **2005**, *17*, 1–9. [CrossRef]

34. FINLEX ®-Viranomaisten Määräyskokoelmat: Maa-ja Metsätalousministeriö-01.09.2011 1784/14/2011. n.d. Available online: http://www.finlex.fi/fi/viranomaiset/normi/400001/37638 (accessed on 1 September 2020).

35. Bekendtgørelse om Anvendelse af Bioaske til Jordbrugsformål (Bioaskebekendtgørelsen)-Retsinformation.dk. n.d. Available online: https://www.retsinformation.dk/forms/r0710.aspx?id=116609 (accessed on 1 September 2020).

36. Van Dijen, F. Biofficiency. Ash Utilization-Draft for Technical Regulations (D6.3). Ref.Ares(2019)645078518/10/2019. Available online: https://ec.europa.eu/research/participants/documents/downloadPublic?documentIds=080166e5c82e4e2f&appId=PPGMS (accessed on 1 September 2020).

37. Regulation (EU) 2019/1009 of the European Parliament and of the Council of 5 June 2019 Laying down Rules on the Making Available on the Market of EU Fertilising Products and Amending Regulations (EC) No. 1069/2009 and (EC) No. 1107/2009 and repealing Regulation (EC) No 2003/2003. Available online: https://eur-lex.europa.eu/legal-content/EN/TXT/?uri=CELEX%3A32019R1009 (accessed on 1 September 2020).

38. Regulation (EC) No. 2003/2003 of the European Parliament and of the Council of 13 October 2003 Relating to Fertilisers. Available online: https://eur-lex.europa.eu/legal-content/EN/TXT/PDF/?uri=CELEX:32003R2003&from=EN (accessed on 1 September 2020).

39. European Committee for Standardization. *EN 196-2:2013. Method of Testing Cement. Chemical Analysis of Cement*; European Committee for Standardization (CEN) Standards: Brussels, Belgium, 2013.

40. Polish Committee for Standardization. *Standard PN EN 13657: 2006, Characterization of Waste-Digestion for Subsequent Determination of Aqua Regia Soluble Portion of Elements*; Polish Committee for Standardization: Warsaw, Poland, 2006.

41. Polish Committee for Standardization. *Standard PN EN ISO 17294-2: 2016-11, Water Quality-Application of Inductively Coupled Plasma Mass Spectrometry (ICP-MS)-Part 2: Determination of selected elements including Uranium Isotopes (ISO 17294-2:2016)*; Polish Committee for Standardization: Warsaw, Poland, 2006.

42. Polish Committee for Standarization. *Standard PN-EN 15407:2011, Solid Recovered Fuels-Methods for the Determination of Carbon (C), Hydrogen (H) and Nitrogen (N) Content*; Polish Committee for Standardization: Warsaw, Poland, 2011.

43. Polish Committee for Standarization. *Standard PN-EN 15408:2011, Solid Recovered Fuels-Methods for the Determination of Sulphur (S), Chlorine (Cl), Fluorine (F) and Bromine (Br) content*; Polish Committee for Standardization: Warsaw, Poland, 2011.

44. Polish Committee for Standardization. *Standard PN-EN 12457:2006. Characterisation of Waste. Leaching. Compliance Test for Leaching of Granular Waste Materials and Sludges. Part 1–4*; Polish Committee for Standardization: Warsaw, Poland, 2006.

45. Polish Committee for Standardization. *Standard EN ISO 10304-1:2009/AC:2012 Water Quality—Determination of Dissolved Anions by Liquid Chromatography of Ions—Part 1: Determination of Bromide, Chloride, Fluoride, Nitrate, Nitrite, Phosphate and Sulfate—Technical Corrigendum 1 (ISO 10304-1:2007/Cor 1:2010)*; Polish Committee for Standardization: Warsaw, Poland, 2012.

46. Polish Committee for Standardization. *Standard PN EN ISO 14911:2002. Water Quality-Determination of Dissolved L[I+], N[A+], NH[4 +], K[+], MN[2+], CA[2+], MG[2+], SR[2+] AND BA[2+] Using Ion Chromatography-Method for Water and Waste Water*; Polish Committee for Standardization: Warsaw, Poland, 2002.

47. Polish Committee for Standardization. *Standard PN-R-04023:1996. Chemical and Agricultural Analysis of soil—Determination of Available Phosphorus Content in Mineral Soils*; Polish Committee for Standardization: Warsaw, Poland, 1996.

48. Polish Committee for Standardization. *Standard PN-R-04023:1996/A1:2002. Chemical and Agricultural Analysis of Soil—Determination of Available Potassium in Mineral Soils*; Polish Committee for Standardization: Warsaw, Poland, 2002.

49. Tessier, A.; Campbell, P.G.C.; Bisson, M. Sequential extraction procedure for the speciation of particulate trace metals. *Anal. Chem.* **1979**, *51*, 844–851. [CrossRef]

50. Ure, A.M.; Quevauviller, P.; Muntau, H.; Griepink, B. Speciation of Heavy Metals in Soils and Sediments. An Account of the Improvement and Harmonization of Extraction Techniques Undertaken Under the Auspices of the BCR of the Commission of the European Communities. *Int. J. Environ. Anal. Chem.* **1993**, *51*, 135–151. [CrossRef]

51. Polish Committee for Standardization. *Standard PN EN ISO 17294-1: 2007-11, Water Quality—Application of Inductively Coupled Plasma Mass Spectrometry (ICP-MS)-Part 1: General guidelines*; Polish Committee for Standardization: Warsaw, Poland, 2007.

52. Gariglio, N.F.; Buyatti, M.A.; Pilatti, R.A.; Russia, D.E.G.; Acosta, M.R. Use of a germination bioassay to test compost maturity of willow (*Salix* sp.) sawdust. *N. Z. J. Crop. Hortic. Sci.* **2002**, *30*, 135–139. [CrossRef]

53. Ciesielczuk, T.; Rosik-Dulewska, C.; Poluszyńska, J.; Miłek, D.; Szewczyk, A.; Sławińska, I. Acute Toxicity of Experimental Fertilizers Made of Spent Coffee Grounds. *Waste Biomass Valorization* **2017**, *9*, 2157–2164. [CrossRef]

54. Teacă, C.A.; Bodîrlău, R. Assessment of toxicity of industrial waste crop plant assays. *BioResources* **2008**, *3*, 1130–1145.

55. Oleszczuk, P.; Hollert, H. Comparison of sewage sludge toxicity to plants and invertebrates in three different soils. *Chemosphere* **2011**, *83*, 502–509. [CrossRef]

56. Baran, A.; Tarnawski, M. Phytotoxkit/Phytotestkit and Microtox® as tools for toxicity assessment of sediments. *Ecotoxicol. Environ. Saf.* **2013**, *98*, 19–27. [CrossRef]

57. Polish Committee for Standardization. *Standard PN EN ISO 11269-1: 2013, Soil Quality—Determination of the Effects of Pollutants on Soil Flora–Part 1: Method for Measurement of Inhibition of Root Growth (ISO 11269-1:2012)*; Polish Committee for Standardization: Warsaw, Poland, 2013.

58. Uliasz-Bocheńczyk, A.; Pawluk, A.; Sierka, J. Wymywalności zanieczyszczeń z popiołów lotnych ze spalania biomasy. *Miner. Resour. Manag.* **2015**, *31*, 145–156.

59. Dz.U. *119.765. Regulation of the Minister of Agriculture and Rural Development of June 18, 2008 on the Implementation of Certain Provisions of the Act on Fertilizers and Fertilization*; Journal of Laws of the Republic of Poland: Warsaw, Poland, 2008.

60. Saletnik, B.; Zaguła, G.; Bajcar, M.; Czernicka, M.; Puchalski, C. Biochar and biomass ash as a soil ameliorant: The effect on selected soil properties and yield of Giant Miscanthus (Miscanthus × giganteus). *Energies* **2018**, *11*, 2535. [CrossRef]

61. Zapałowska, A.; Puchalski, C.; Hury, G.; Makarewicz, A. Influence of fertilization with the use of biomass ash and sewage sludge on the chemical coposition of Jerusalem Artichole used for energy related purposes. *J. Ecol. Eng.* **2017**, *18*, 235–245. [CrossRef]

62. Vassilev, S.V.; Baxter, D.; Andersen, L.K.; Vassileva, C.G.; Morgan, T.J. An overview of the organic and inorganic phase composition of biomass. *Fuel* **2012**, *94*, 1–33. [CrossRef]

63. Cobreros, C.; Reyes-Araiza, J.L.; Manzano-Ramírez, A.; Nava, R.; Rodríguez, M.; Mondragón-Figueroa, M.; Apátiga, L.M.; Rivera-Muñoz, E.M. Barley straw ash: Pozzolanic activity and comparison with other natural and artificial pozzolans from Mexico. *BioResources* **2015**, *10*, 3757–3774. [CrossRef]

64. Jaworek, A.; Czech, T.; Sobczyk, A.T.; Krupa, A. Properties of biomass vs. coal fly ashes deposited in electrostatic precipitator. *J. Electrost.* **2013**, *71*, 165–175. [CrossRef]

65. García-Jiménez, A.; Trejo-Téllez, L.I.; Guillén-Sánchez, D.; Gómez-Merino, F.C. Vanadium stimulates pepper plant growth and flowering, increases concentrations of amino acids, sugars and chlorophylls, and modifies nutrient concentrations. *PLoS ONE* **2018**, *13*, e0201908. [CrossRef] [PubMed]

66. Wang, J.F.; Liu, Z. Effect of vanadium on the growth of soybean seedlings. *Plant Soil* **1999**, *216*, 47–51. [CrossRef]

67. Shanker, A.; Djanaguiraman, M.; Venkateswarlu, B. Chromium interactions in plants: Current status and future strategies. *Metallomics* **2009**, *1*, 375–383. [CrossRef]

68. Dz.U. *2016 poz. 1395. Regulation of the Minister of the Environment of September 1, 2016 on the Method of Assessing the Pollution of the Earth's Surface*; Journal of Laws of the Republic of Poland: Warsaw, Poland, 2016.

69. Badura, L. Metale w glebach-stan zagrożenia na przykładzie województwa katowickiego. In *Ekologiczne problemy Górnego Śląska, red J. Śliwiok. Wyd*; Wszechnica Górnośląskiego Towarzystwa Przyjaciół Nauk im. W. Roździeńskiego: Katowice, Poland, 1995. (In Polish)

70. Wang, L.K.; Chen, J.P.; Hung, Y.; Shammas, N.K. *Heavy Metals in the Environment*; CRP Press Taylor & Francis Group: Boca Raton, FL, USA, 2009.

71. Abbasi, S.A.; Abbasi, N.; Soni, R. *Heavy Metals in the Environment*; Mittal Publications: New Delhi, India, 2019.

72. Bååth, E. Effects of heavy metals in soil on microbial processes and populations (a review). *Water Air Soil Pollut.* **1989**, *47*, 335–379. [CrossRef]

73. Zhao, F.-J.; Tausz, M.; De Kok, L.J. Role of Sulfur for Plant Production in Agricultural and Natural Ecosystems. In *Sulfur Metabolism in Phototrophic Organisms*; Hell, R., Dahl, C., Knaff, D., Leustek, T., Eds.; Springer: Dordrecht, The Netherlands, 2008; Volume 27, pp. 417–435.

74. Hawkesford, M.J. Plant responses to sulphur deficiency and the genetic manipulation of sulphate transporters to improve S-utilization efficiency. *J. Exp. Bot.* **2000**, *51*, 131–138. [CrossRef]

75. Tavakkoli, E.; Rengasamy, P.; McDonald, G.K. High concentrations of $Na^+$ and $Cl^-$ ions in soil solution have simultaneous detrimental effects on growth of faba bean under salinity stress. *J. Exp. Bot.* **2010**, *61*, 4449–4459. [CrossRef]

76. Ohno, T. Erich MS Effect of wood ash application on soil pH and soil test nutrient levels. *Agric. Ecosyst. Environ.* **1990**, *32*, 223–239. [CrossRef]

77. Pels, J.R.; De Nie, D.S.; Kiel, J.H.A. Utilization of ashes from biomass combustion and gasification. In Proceedings of the 14th European Biomass Conference & Exhibition, Paris, France, 17–21 October 2005. ECN-RX-05-182.

78. Sarkar, S.K.; Favas, P.J.C.; Rakshit, D.; Satpathy, K.K. Geochemical speciation and risk assessment of heavymetals in soils and sediments. In *Environmental Risk Assessment of Soil Contamination*; Hernandez-Soriano, M.C., Ed.; InTech: London, UK, 2014; p. 918.

79. Agrawal, R.; Kumar, B.; Priyanka, K.; Narayan, C.; Shukla, K.; Sarkar, J. Anshumali Micronutrient Fractionation in Coal Mine-Affected Agricultural Soils, India. *Bull. Environ. Contam. Toxicol.* **2016**, *96*, 449–457. [CrossRef] [PubMed]

80. Dong, D. Adsorption of Pb and Cd onto metal oxides and organic material in natural surface coatings as determined by selective extractions: New evidence for the importance of Mn and Fe oxides. *Water Res.* **2000**, *34*, 427–436. [CrossRef]

*Article*

# The Effects of Microalgae Biomass Co-Substrate on Biogas Production from the Common Agricultural Biogas Plants Feedstock

Marcin Debowski [1,*], Marta Kisielewska [1], Joanna Kazimierowicz [2], Aleksandra Rudnicka [1], Magda Dudek [1], Zdzisława Romanowska-Duda [3] and Marcin Zieliński [1]

[1]   Department of Environmental Engineering, Faculty of Geoengineering, University of Warmia and Mazury in Olsztyn, 10-719 Olsztyn, Poland; jedrzejewska@uwm.edu.pl (M.K.); aleksandra.rudnicka@uwm.edu.pl (A.R.); magda.dudek@uwm.edu.pl (M.D.); marcin.zielinski@uwm.edu.pl (M.Z.)

[2]   Department of Water Supply and Sewage Systems, Faculty of Civil Engineering and Environmental Sciences, Bialystok University of Technology, 15-351 Białystok, Poland; j.kazimierowicz@pb.edu.pl

[3]   Department of Plant Ecophysiology, Faculty of Biology and Environmental Protection, University of Lodz, Banacha St. 12/13, 90-237 Lodz, Poland; zdzislawa.romanowska@biol.uni.lodz.pl

*   Correspondence: marcin.debowski@uwm.edu.pl

Received: 23 March 2020; Accepted: 23 April 2020; Published: 1 May 2020

**Abstract:** The aim of this study was to determine the effects on methane production of the addition of microalgae biomass of *Arthrospira platensis* and *Platymonas subcordiformis* to the common feedstock used in agricultural biogas plants (cattle manure, maize silage). Anaerobic biodegradability tests were carried out using respirometric reactors operated at an initial organic loading rate of 5.0 kg volatile solids (VS)/$m^3$, temperature of 35°C, and a retention time of 20 days. A systematic increase in the biogas production efficiency was found, where the ratio of microalgae biomass in the feedstock increased from 0% to 40% (%VS). Higher microalgae biomass ratio did not have a significant impact on improving the efficiency of biogas production, and the biogas production remained at a level comparable with 40% share of microalgae biomass in the feedstock. This was probably related to the carbon to nitrogen (C/N) ratio decrease in the mixture of substrates. The use of *Platymonas subcordiformis* ensured higher biogas production, with the maximum value of 1058.8 ± 25.2 L/kg VS. The highest content of methane, at an average concentration of 65.6% in the biogas produced, was observed in setups with *Arthrospira plantensis* biomass added at a concentration of between 20%–40% to the feedstock mixture.

**Keywords:** microalgae; anaerobic digestion; biogas; respirometric reactors

## 1. Introduction

Biomass is currently regarded as one of the most important sources of renewable energy that will allows the global energy goals to be met [1]. Today biomass represents nearly 8% of the total primary energy supply in Europe [2]. The main conversion pathway for converting biomass to bioenergy carriers is anaerobic digestion (AD) [3]. During AD biogas is produced, which is a renewable energy source that can be used for the production of electricity, heat, or in vehicle transportation [4]. At present, the biomass used in agricultural biogas plants is mainly terrestrial plants [5–7], whose an intensive cultivation may negatively affect the global supply of food and feed [8]. Thus, there is a need to search for alternative sources of biomass to replace food feedstocks.

Previous studies indicate that microalgae biomass has a potential for use as an organic substrate for bioenergy production. [9]. Microalgae biomass for biogas production can be obtained from closed photobioreactors, open ponds, and from natural water reservoirs [10]. Previous reports indicate that the biomass of *Scenedesmus* sp. [11], *Spirulina* sp. [12,13], *Euglena* sp. and *Chlorella vulgaris* [14], *Melosira* sp. and *Oscillatoria* sp. [15], as well as the benthic multicellular algae including *Laminaria* sp., *Macrocystis* sp. [16], *Gracilaria ceae* [17], *Ulva* sp. [18] and *Macrosystis pyrifera, Tetraselmis, Gracilaria tikvahiae*, and *Hypnea* sp. [19,20] are good sources to produce biomethane.

Microalgae biomass has many advantages over conventional energy crops. Microalgae accumulate large amounts of polysaccharides and lipids in their cells, and are deprived of hardly degradable lignocellulosic compounds [21]. They are characterized by a high growth rate and do not compete with crops for nutritional and feed purposes [22,23]. Thus, algae biomass offers great potential as a resource for the production of various energy carriers, such as biohydrogen, bioethanol, biodiesel, and biogas [24,25]. The operating problems in anaerobic digestion of algae biomass are associated with the biochemical composition of biomass, where high protein concentration reduces the value of the C/N ratio. However, it can be effectively corrected by co-digestion of algal biomass with feedstock rich in carbon compounds [11].

The combined treatment of several substrates in AD may improve the efficiency of biogas production comparing the yields achieve for each substrate separately. This is due to the positive synergistic effects establish in the digestion feedstock [26,27]. In this way, many missing microelements and nutrients necessary for anaerobic microflora are supplied to the reactor [28]. Additional benefits associated with co-digestion of the selected substrates may also relate to other factors, such as technological, economic and environmental aspects [29,30]. Finally, the increasing interest in developing microalgae-to-biofuel technology requires a detailed assessment of technological parameters of AD with a process optimization.

The aim of this research was to investigate the potential of *Arthrospira platensis* and *Platymonas subcordiformis* microalgae biomass as the feedstock for anaerobic co-digestion with the common feedstock of agricultural biogas plants, i.e., maize silage and cattle manure, to enhance biogas/methane yield.

## 2. Materials and Methods

### 2.1. Feedstock Origin and Characteristics

The microalgal biomass used in this study was collected from our own culture. The two vertical and tubular photobioreactors made of transparent plexiglass were used for separate cultivation of *Arthrospira platensis* and *Platymonas subcordiformis*. The working volume of each reactor was 50 L (inner diameter 200 mm, height 1700 mm). The light was provided with white reflectors (700 lux, Osram, Germany). The algal biomass was cultivated for 15 days. After the cultivation process was ended, the microalgae biomass was harvested, and then dehydrated by preliminary sedimentation followed by centrifugation (3000 rpm for 6 min). Dehydrated biomass was later mixed with other substrates (i.e., cattle slurry and maize silage).

Substrates for AD (cattle slurry, maize silage) originated from the Research Station of University of Warmia and Mazury in Olsztyn in Bałdy (Poland). Samples of substrates were collected in 5 kg amounts from five different places in storage fields; 1 kg from each place. They were subsequently mixed in order to obtain a homogenous sample of cattle slurry and sample of maize silage.

In the study, the substrates selected were the model organic substrates of maize silage and cattle slurry commonly used in agricultural biogas plants, as well as microalgae species characterized by high growth rate, which is an important factor for industrial applications. The characteristics of the feedstock substrates used in the study are presented in Table 1.

**Table 1.** Characteristics of organic substrates used for the feedstock preparation. TN: total nitrogen; TP: total phosphorus; TC: total carbon; TOC: total organic carbon; C/N: carbon to nitrogen.

| Parameter | Unit | Maize Silage | Cattle Slurry | *Arthrospira Platensis* | *Platymonas Subcordiformis* |
|---|---|---|---|---|---|
| Total solids (TS) | (% fresh mass) | 30.2 ± 0.9 | 9.5 ± 1.2 | 7.2 ± 1.0 | 8.4 ± 0.6 |
| Volatile solids | (% TS) | 93.8 ± 0.2 | 74.9 ± 0.6 | 91.5 ± 0.9 | 87.1 ± 0.9 |
| Mineral solids | (% TS) | 6.2 ± 1.3 | 25.1 ± 1.3 | 8.5 ± 0.9 | 12.9 ± 0.9 |
| TN | (g/kg TS) | 11.1 ± 0.9 | 49.8 ± 3.7 | 58.1 ± 5.7 | 43.4 ± 1.7 |
| TP | (g/kg TS) | 2.4 ± 0.3 | 22.4 ± 1.2 | 10.3 ± 1.0 | 19.9 ± 1.3 |
| TC | (g/kg TS) | 460.1 ± 12.9 | 390.8 ± 17.4 | 493.4 ± 17.1 | 474.8 ± 11.5 |
| TOC | (g/kg TS) | 441.0 ± 15.1 | 320.1 ± 13.9 | 434.3 ± 12.7 | 439.4 ± 27.3 |
| C/N | - | 39.6 ± 1.7 | 7.9 ± 0.6 | 8.5 ± 0.5 | 10.9 ± 0.4 |
| pH | - | 7.7 ± 0.1 | 7.1 ± 0.1 | 8.1 ± 0.1 | 7.9 ± 0.3 |

## 2.2. Experimental Setup

Two different experimental series were performed, where either *Arthrospira plantensis* (series 1) or *Platymonas subcordiformis* (series 2) was added as algal biomass, and the feedstock was investigated in batch AD assays. In each series six different setups, based on the different composition of the substrate mixtures added, were investigated (Table 2). The characteristics of the different substrate mixtures used in the batch AD assays are presented in Table 3.

**Table 2.** Experimental setup. VS: volatile solids.

| | Concentration of Individual Substrates (% VS) | | | |
|---|---|---|---|---|
| | Setup | Maize silage | Cattle slurry | Arthrospira platensis |
| | 1 | 70 | 30 | 0 |
| | 2 | 67 | 23 | 10 |
| Series 1 | 3 | 60 | 20 | 20 |
| | 4 | 45 | 15 | 40 |
| | 5 | 30 | 10 | 60 |
| | 6 | 15 | 5 | 80 |
| | Setup | Maize silage | Cattle slurry | Platymonas subcordiformis |
| | 1 | 70 | 30 | 0 |
| | 2 | 67 | 23 | 10 |
| Series 2 | 3 | 60 | 20 | 20 |
| | 4 | 45 | 15 | 40 |
| | 5 | 30 | 10 | 60 |
| | 6 | 15 | 5 | 80 |

**Table 3.** Characteristics of substrates mixtures used as the feedstocks for anaerobic digestion in experimental setups (S).

| Parameter | Unit | Series 1 | | | | | | Series 2 | | | | | |
|---|---|---|---|---|---|---|---|---|---|---|---|---|---|
| | | S1 | S2 | S3 | S4 | S5 | S6 | S1 | S2 | S3 | S4 | S5 | S6 |
| Total solids | (% fresh mass) | $24.0 \pm 0.9$ | $23.2 \pm 0.9$ | $21.5 \pm 0.9$ | $17.9 \pm 0.9$ | $14.3 \pm 0.9$ | $10.8 \pm 0.9$ | $24.0 \pm 0.9$ | $23.3 \pm 0.9$ | $21.7 \pm 0.9$ | $18.4 \pm 0.8$ | $15.1 \pm 0.7$ | $11.8 \pm 0.7$ |
| Volatile solids | (% TS) | $88.1 \pm 0.3$ | $89.2 \pm 0.4$ | $89.6 \pm 0.4$ | $90.0 \pm 0.6$ | $90.5 \pm 0.7$ | $90.9 \pm 0.8$ | $88.1 \pm 0.3$ | $88.8 \pm 0.4$ | $88.7 \pm 0.4$ | $88.3 \pm 0.6$ | $87.9 \pm 0.1$ | $87.5 \pm 0.8$ |
| Mineral solids | (% TS) | $11.9 \pm 1.3$ | $10.8 \pm 1.3$ | $10.4 \pm 1.3$ | $9.9 \pm 1.2$ | $9.5 \pm 1.1$ | $9.0 \pm 1.0$ | $11.9 \pm 1.3$ | $11.2 \pm 1.3$ | $11.3 \pm 1.3$ | $11.7 \pm 1.2$ | $12.1 \pm 1.1$ | $12.5 \pm 1.0$ |
| TN | (g/kg TS) | $22.7 \pm 1.7$ | $24.7 \pm 2.0$ | $28.2 \pm 2.4$ | $35.7 \pm 3.2$ | $43.2 \pm 4.0$ | $50.6 \pm 4.9$ | $22.7 \pm 1.7$ | $23.2 \pm 1.6$ | $25.3 \pm 1.6$ | $29.8 \pm 1.7$ | $34.3 \pm 1.7$ | $38.9 \pm 1.7$ |
| TP | (g/kg TS) | $8.4 \pm 0.5$ | $7.8 \pm 0.5$ | $7.9 \pm 0.6$ | $8.6 \pm 0.7$ | $9.2 \pm 0.8$ | $9.7 \pm 0.9$ | $8.4 \pm 0.5$ | $8.8 \pm 0.6$ | $9.9 \pm 0.7$ | $12.4 \pm 0.8$ | $14.9 \pm 1.0$ | $17.5 \pm 1.2$ |
| TC | (g/kg TS) | $439.3 \pm 14.3$ | $447.5 \pm 14.4$ | $452.9 \pm 14.7$ | $463.0 \pm 15.3$ | $473.2 \pm 15.9$ | $483.3 \pm 16.5$ | $439.3 \pm 14.3$ | $445.7 \pm 13.8$ | $449.2 \pm 13.6$ | $455.6 \pm 13.0$ | $462.0 \pm 12.5$ | $468.4 \pm 12.0$ |
| TOC | (g/kg TS) | $404.8 \pm 14.7$ | $412.6 \pm 14.6$ | $415.5 \pm 14.4$ | $420.2 \pm 13.9$ | $424.9 \pm 13.6$ | $429.6 \pm 13.1$ | $404.8 \pm 14.7$ | $413.1 \pm 16.0$ | $416.5 \pm 17.3$ | $422.3 \pm 19.8$ | $427.9 \pm 22.3$ | $433.7 \pm 24.8$ |
| C/N | - | $30.1 \pm 1.3$ | $29.2 \pm 1.3$ | $27.1 \pm 1.2$ | $22.4 \pm 1.0$ | $17.8 \pm 0.9$ | $13.1 \pm 0.7$ | $30.1 \pm 1.3$ | $29.5 \pm 1.3$ | $27.5 \pm 1.2$ | $23.4 \pm 1.0$ | $19.2 \pm 0.8$ | $15.1 \pm 0.6$ |
| pH | - | $7.6 \pm 0.1$ | $7.6 \pm 0.1$ | $7.7 \pm 0.1$ | $7.8 \pm 0.1$ | $7.9 \pm 0.1$ | $8.0 \pm 0.1$ | $7.6 \pm 0.1$ | $7.6 \pm 0.1$ | $7.7 \pm 0.1$ | $7.7 \pm 0.2$ | $7.8 \pm 0.2$ | $7.9 \pm 0.3$ |

### 2.3. Batch Anaerobic Digestion Assays

Batch anaerobic digestion assays were carried out with respirometers (WTW, Germany) that consisted of bottles with reaction chamber volume of 0.5 L and measuring heads as the pressure sensors. The pressure increasing in the bottles caused by biogas production was measured and recorded every 180 min.

The bottles were filled with anaerobic inoculum to the volume of 200 mL and the feedstock mixture to a volume that ensured the set organic loading rate (OLR). The inoculum was taken from the closed fermentation chamber of municipal wastewater treatment plant operating at OLR of 2.0 kg volatile solids (VS)/m$^3$·d, hydraulic retention time (HRT) of 20 days and under mesophilic conditions of 35 °C. The anaerobic inoculum characteristic is shown in Table 4. The mixture volume of inoculum and feedstock in the bottles ensured an initial OLR of 5.0 g VS/L. At the beginning of assays, anaerobic conditions inside the respirometers were obtained by purging nitrogen gas to remove atmospheric air. Batch AD assays were carried out for a period of 20 days and at a constant temperature of 35 °C ± 0.5 °C.

**Table 4.** Characteristic of anaerobic inoculum for batch anaerobic digestion (AD) assays

| Parameter | Unit | Value |
| --- | --- | --- |
| Total solids | (% fresh mass) | 3.8 ± 0.2 |
| Volatile solids | (% TS) | 68.5 ± 2.5 |
| Mineral solids | (% TS) | 31.5 ± 2.4 |
| TN | (g/kg TS) | 33.1 ± 3.4 |
| TP | (g/kg TS) | 1.7 ± 0.2 |
| TC | (g/kg TS) | 309.1 ± 28.4 |
| TOC | (g/kg TS) | 199.4 ± 34.3 |
| C/N | - | 9.3 ± 0.1 |
| pH | - | 7.2 ± 0.3 |

For the determination of biogas potential the ideal gas law was used, and the pressure changes inside the bottles were converted to the biogas volumes produced under normal conditions. The biogas production rate (r) was determined for each experimental setup. The non-linear regression and iterative method were used to determine reaction rate constants (k), (Statistica 13.1 PL software). In the iterative method, at each iterative step, the function is replaced with the linear differential for the designated parameters. The curve fitting test ($\varphi^2$ coefficient) was performed to find the best fit of designated parameters to the experimental data points. It was assumed that the model was adapted to the experimental data when $\varphi^2$ value did not exceed 0.2.

### 2.4. Analytical Methods

The gravimetric method enabled the determination of TS (total solids) and VS (volatile solids) concentrations. The samples of feedstock mixtures and anaerobic inoculum were dried at 105 °C and then determined for the total carbon (TC), total organic carbon (TOC) and total nitrogen (TN) concentrations by Flash 2000 analyzer (Thermo Fisher Scientific Inc.). The concentrations of total phosphorus (TP) were measured with a spectrophotometer DR 2800 with mineralizer (HACH Lange, Germany). The aqueous solution for pH determination was prepared by weighing 10 g of the homogenized air-dried sample in a 100 mL glass beaker, and then adding 50 mL distilled water and mixing.

The biogas composition ($CH_4$, $CO_2$, $O_2$, $H_2$, $H_2S$ and $NH_3$) was analyzed every 24 h using gas chromatography (GC). A gastight syringe was used to inject gas sample volume of 20 mL into a gas chromatograph (GC, 7890A Agilent) equipped with a thermal conductivity detector (TCD). For separation of gases, the two Hayesep Q columns (80/100 mesh), two molecular sieve columns

(60/80 mesh), and a Porapak Q column (80/100) operating at a temperature of 70 °C were used. The operational temperatures of injection and detector ports were respectively 150 °C and 250 °C. Helium and argon were applied as the carrier gases, both at the flow of 15 mL/min. The biogas composition was additionally evaluated using a GMF 430 analyzer (Gas Data).

*2.5. Statistical Methods*

The data obtained in the study were statistically processed by using Statictica 13.1 PL package (StatSoft, Inc.). The W Shapiro–Wilk's test was used to see if variables were normally distributed. One way analysis of variance (ANOVA) was used to determine whether there were any statistically significant differences between the means. The dependent variables were the amount of biogas and the methane content in biogas, while the grouping variable was the feedstock composition. The relationship between the different composition of the feedstock was determined using Pearson's correlation. The Levene's test was used to determine if the comparing groups had equal variances. The Tukey's HSD (honest significant difference) test was used to examine the significance of differences between the analyzed variables. The differences were considered significant at p = 0.05. To assess the biogas components depending on the feedstock characteristic, the F test and t test were used. The significance level was 0.01 for F test and 0.025 for t.

A stepwise regression was used to find the best multiple regression model with only statistically significant predictors from a set of potential predictive variables. The predictors with significant impact on changes in the biogas production B (L/kgVS) in models were TN (g/kg TS) and VS (%TS). The fit of the models to the empirical data was assessed using determination coefficients. The significance of polynomial regression models was verified using F-statistic and reference to the critical values. Lack-of-fit test was performed to check if the proposed statistical models fitted well. The test involved comparing the proposed models with models containing the remainder of the explanatory variables omitted in the proposed models. The models were subjected to the estimation tests. Examination of residuals to check for the model and the accuracy of assumptions was assessed. The assumption of normality of residuals distribution was verified and the correctness of models was assessed by plotting the value of residuals against predicted values (Statistica 13.1 PL).

## 3. Results and Discussion

The studies revealed that mixing the microalgae biomass belonging to *Arthrospira platensis* and *Platymonas subcordiformis* species and the biogas plant feedstock (cattle slurry and maize silage) caused improvements to to biogas yield and composition. In the study, the biogas and methane yields coming from the mixture of maize silage and cattle slurry achieved respectively $620.5 \pm 14.6$ $L_{biogas}$/kgVS and $343.1 \pm 16.4$ $L_{CH4}$/kgVS. The addition of the *Arthrospira platensis* biomass (up to a concentration of 10%) enhanced biogas production to $714.4 \pm 16.1$ $L_{biogas}$/kgVS while the addition of 80% resulted in $923.6 \pm 25.1$ $L_{biogas}$/kgVS. The methane yield also increased from $390.1 \pm 11.8$ $L_{CH4}$/kgVS (10% of microalgal biomass) to $581.0 \pm 24.5$ $L_{CH4}$/kgVS (40% of microalgal biomass). When *Platymonas subcordiformis* biomass was tested, the biogas and methane yields ranged from $918.0 \pm 23.6$ $L_{biogas}$/kgVS and $487.5 \pm 19.6$ $L_{CH4}$/kgVS, respectively (for 10% of microalgal biomass) to $1058.8 \pm 25.2$ $L_{biogas}$/kgVS and $577.1 \pm 24.3$ $L_{CH4}$/kgVS, respectively (for 80% of microalgal biomass).

Giuliano et al. studied co-digestion of energy crops and cattle manure [31]. Biogas production obtained varied from 320 to 370 $L_{biogas}$/kgVS$_{fed}$ in mesophilic conditions. In turn, Amon et al. (2007) achieved the methane production from maize and dairy cattle manure in the range of 312–365 $L_{CH4}$/kgVS (milk ripeness) and 268–286 $L_{CH4}$/kgVS (full ripeness) [32]. Kalamaras and Kotsopoulos found the methane potential of 267 $L_{CH4}$/kgVS from the same substrate co-digestion [33]. The higher efficiencies of biogas production during co-digestion of algae biomass and others organic feedstocks are attributed to the synergistic effects established in anaerobic reactors. In anaerobic digestion of mixed organic substrates, algae biomass is a source of nitrogen and microelements for the growth of microorganisms. This has been confirmed by the studies of others authors [27]. Similar conclusions have also been made by Matsui et al. [34], who operated a pilot-scale reactor where macroalgae of Laminaria sp. and Ulva sp. were mixed with others organic waste feedstocks.

In both series of the experiment, the maximum biogas production was observed in setups with microalgae content in feedstock ranged from 40% to 80% (%VS). In series 1, the highest biogas production was within the range of 885.7 ± 20.2 L/kg VS - 923.6 ± 25.1 L/kg VS, while the rate of reaction varied from r = 392 mL/d to r = 426 mL/d (Table 5, Figure 1). In turn, in series 2, the results oscillated between 1012.0 ± 24.1 mL/kg VS and 1058.8 ± 25.2 mL/kg VS with the rate from r = 512 mL/d to r = 560 mL/d (Table 5, Figure 1). It was significantly higher ($p < 0.05$) than in series 1. The methane content in biogas of series 1 averaged: 65.6 ± 1.3% in setup 4, 57.0 ± 1.8% in setup 5 and 53.4 ± 0.8% in setup 6. In series 2 it was 52.9 ± 1.05% in setup 4, 54.5 ± 1.08% in setup 5 and 54.5 ± 0.98% in setup 6 (Table 6). Significantly lower biogas production of 620.49 ± 14.55 L/kg VS ($p < 0.05$) was noted in setup 1, where the feedstock for anaerobic digestion consisted only of maize silage and cattle slurry (Figure 1, Table 5). The methane content in biogas obtained in setup 1 averaged 55.29 ± 1.32% (Table 6).

Others authors [35] have indicated that the potential of biogas production depends directly on microalgae species. However, no correlation was found between the taxonomic group of alage and the process efficiency in the experiments with six phytoplankton species (*Chlamydomonas reinwardtii*, *Dunaliella salina* and *Scenedesmus obliquus* of the class *Chlorophyceae*, *Chlorella kessleri* of the class *Trebouxiophyceae*, *Euglena gracilis* of the class *Euglenoidea* and cyanobacteria *Arthrospira platensis* of the class *Cyanophyceae*). The biogas production obtained from *Chlamydomonas reinhardtii* reached 587 ± 8.8 L/kg VS, while the biomass of *Dunaliella salina* achieved 505 ± 24.8 L/kg VS. Anaerobic digestion of cyanobacteria *Arthrospira platensis* and *Euglena gracilis* resulted in a lower biogas production, which was 481 ± 13.8 L/kg VS and 485 ± 3.0 L/kg VS respectively. The biogas production from *Chlorella kessleri* and *Scenedesmus obliquus* biomass was the lowest, and attained 335 ± 7.8 L/kg VS and 287 ± 10.1 L/kg VS, respectively [35]. Singh and Gu [36] and Parmar et al. [37] emphasized the impact of the algal species on biogas production efficiency.

The necessity of selecting the appropriate proportions of co-substrates in the feedstock mixture results from the fact that an improper C/N ratio may limit (or even completely inhibit) the growth of anaerobic microflora in AD [14]. Feedstock based on terrestrial energy crops is characterized by a high C/N ratio. Elser et al. (2000) determined the C/N ratio in terrestrial plants to be 36.0 [38]. In turn, the C/N ratio of maize mixture achieved the value of 33.6 and for giant cane mixture it was 35.3 [39]. The C/N ratio ranging from 32.6 to 44.5 was found in maize silage [40]. On the other hand, the feedstock consisted only of microalgae biomass has low C/N ratio (about 10) [41]. Decreasing biogas production in low C/N ratio is attributed to the high concentration of ammonia nitrogen and volatile fatty acids in the chamber of anaerobic reactors. That may cause the inhibition of biochemical pathways [41]. The way to reduce this effect is to mix the organic substrates in appropriate proportions [29]. However, literature review doesn't provide the exact ranges of C/N ratio for undisturbed course of anaerobic digestion. It is assumed that the optimal C/N ratio should be in the range of 16 to 25 [42], although according to others authors it may vary in a wider range from 20 to 70 [43], or even in a narrower range from 12 to 16 [44]. A range of 20 to 30 is also given [45].

**Table 5.** Biogas production in experimental setups (S).

| Setup | Series 1 | | | | | | Series 2 | | | | | |
|---|---|---|---|---|---|---|---|---|---|---|---|---|
| | Biogas | | | Methane | | | Biogas | | | Methane | | |
| | L/kg Fresh Mass | L/kg TS | L/kg VS | L/kg Fresh Mass | L/kg TS | L/kg VS | L/kg Fresh Mass | L/kg TS | L/kg VS | L/kg Fresh Mass | L/kg TS | L/kg VS |
| S1 | 105.5 ± 6.3 | 545.6 ± 13.9 | 620.5 ± 14.6 | 58.3 ± 3.3 | 301.7 ± 13.2 | 343.1 ± 16.4 | 105.5 ± 6.3 | 545.6 ± 13.9 | 620.5 ± 14.6 | 58.3 ± 3.3 | 301.7 ± 13.2 | 343.1 ± 16.4 |
| S2 | 100.0 ± 6.1 | 573.9 ± 14.0 | 714.4 ± 16.1 | 54.6 ± 3.0 | 313.3 ± 14.0 | 390.1 ± 11.8 | 163.2 ± 7.3 | 786.9 ± 18.7 | 918.0 ± 23.6 | 86.7 ± 4.9 | 417.8 ± 14.8 | 487.5 ± 19.6 |
| S3 | 590.7 ± 14.3 | 391.8 ± 11.2 | 775.8 ± 18.2 | 39.2 ± 3.0 | 257.0 ± 13.1 | 508.9 ± 20.3 | 174.6 ± 7.5 | 760.9 ± 18.3 | 926.2 ± 23.1 | 96.2 ± 5.4 | 420.4 ± 14.9 | 510.3 ± 22.8 |
| S4 | 580.7 ± 14.0 | 396.2 ± 11.2 | 885.7 ± 20.2 | 38.5 ± 2.8 | 259.9 ± 13.6 | 581.0 ± 24.5 | 173.5 ± 7.6 | 889.6 ± 19.2 | 1012.0 ± 24.1 | 91.8 ± 5.2 | 470.6 ± 15.5 | 535.3 ± 23.6 |
| S5 | 700.0 ± 15.5 | 459.7 ± 18.6 | 910.2 ± 22.7 | 39.9 ± 3.0 | 262.0 ± 13.8 | 518.8 ± 29.4 | 199.7 ± 8.2 | 810.3 ± 19.0 | 1019.9 ± 23.6 | 108.9 ± 6.4 | 440.3 ± 14.9 | 555.9 ± 24.1 |
| S6 | 110.4 ± 6.1 | 655.9 ± 14.9 | 923.6 ± 25.1 | 58.9 ± 3.5 | 350.2 ± 16.6 | 493.2 ± 20.9 | 226.2 ± 8.7 | 840.4 ± 19.1 | 1058.8 ± 25.2 | 123.3 ± 7.1 | 451.0 ± 15.1 | 577.1 ± 24.3 |

**Table 6.** Biogas composition in experimental setups (S).

| Setup | Series 1 | | | | | | Series 2 | | | | | |
|---|---|---|---|---|---|---|---|---|---|---|---|---|
| | $CH_4$ (%) | $CO_2$ (%) | $O_2$ (%) | $H_2S$ (ppm) | $H_2$ (ppm) | $NH_3$ (ppm) | $CH_4$ (%) | $CO_2$ (%) | $O_2$ (%) | $H_2S$ (ppm) | $H_2$ (ppm) | $NH_3$ (ppm) |
| S1 | 55.3 ± 1.3 | 44.7 ± 1.5 | - | 15 ± 0.9 | 13 ± 0.9 | 10 ± 0.8 | 55.3 ± 1.3 | 44.7 ± 1.5 | - | 15 ± 0.9 | 13 ± 0.9 | 10 ± 0.8 |
| S2 | 54.6 ± 0.4 | 45.4 ± 0.7 | - | 18 ± 0.8 | 20 ± 1.3 | 20 ± 0.9 | 53.1 ± 0.7 | 46.9 ± 1.3 | - | 13 ± 0.9 | 16 ± 1.3 | 20 ± 1.1 |
| S3 | 65.6 ± 1.1 | 43.4 ± 1.6 | - | 17 ± 0.9 | 18 ± 1.4 | 15 ± 1.0 | 55.1 ± 1.1 | 44.9 ± 1.2 | - | 14 ± 0.9 | 10 ± 1.0 | 40 ± 2.0 |
| S4 | 65.6 ± 1.3 | 43.4 ± 1.1 | - | 16 ± 1.1 | 22 ± 1.1 | 15 ± 1.1 | 52.9 ± 1.1 | 47.1 ± 1.2 | - | 10 ± 0.8 | 16 ± 0.9 | 10 ± 1.2 |
| S5 | 57.0 ± 1.8 | 43.0 ± 1.1 | - | 15 ± 1.0 | 21 ± 1.0 | 10 ± 1.0 | 54.5 ± 1.1 | 45.5 ± 1.6 | - | 10 ± 1.0 | 13 ± 0.9 | 18 ± 1.5 |
| S6 | 53.4 ± 0.8 | 46.6 ± 1.0 | - | 14 ± 0.9 | 18 ± 1.0 | 24 ± 1.3 | 54.5 ± 1.0 | 45.5 ± 1.1 | - | 8 ± 0.8 | 13 ± 0.9 | 16 ± 1.5 |

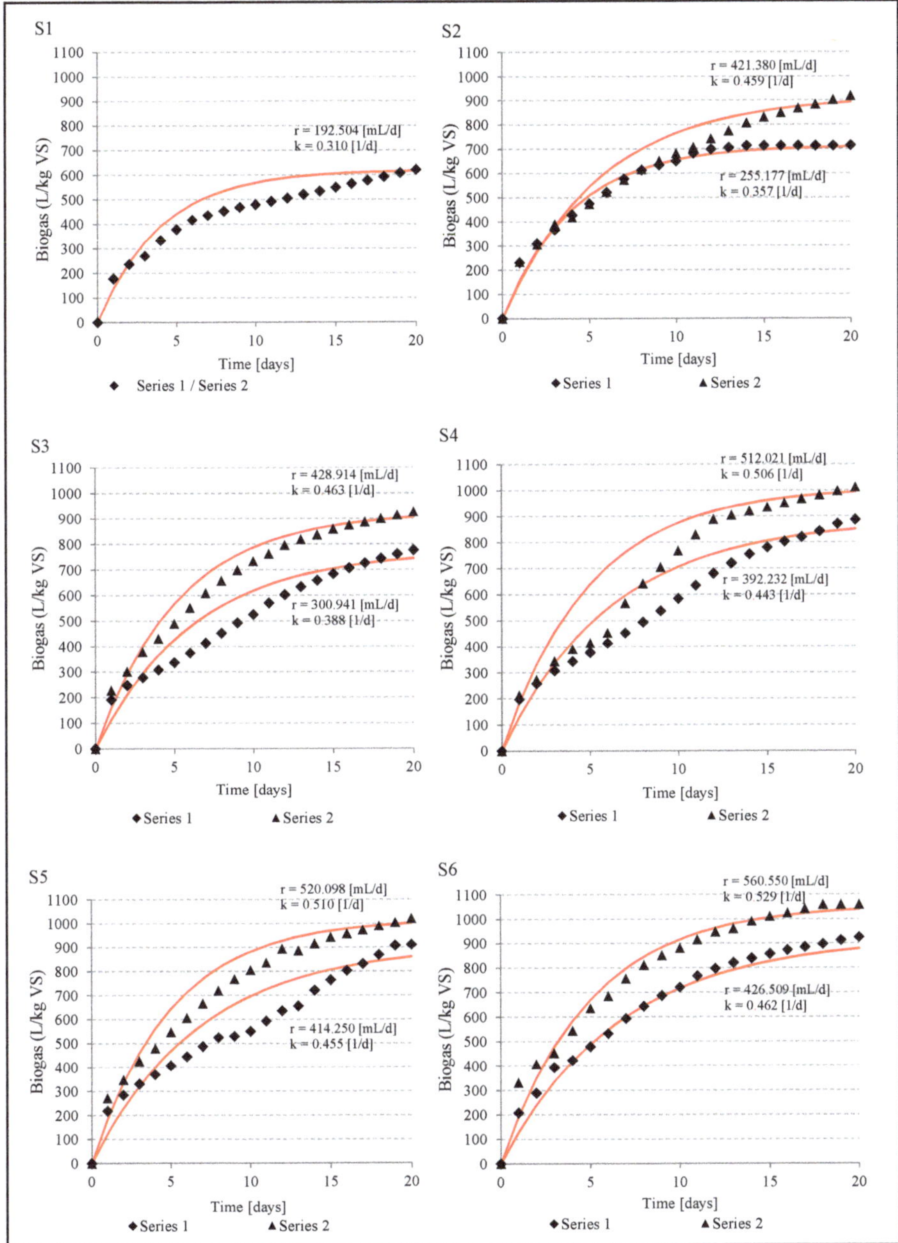

**Figure 1.** Biogas production in batch AD assays over time in experimental setups (S).

In these studies it was found that the presence of microalgae biomass in the feedstock for anaerobic digestion significantly improved the value of the C/N ratio. Nevertheless, the increase in microalgae biomass above 40% of VS content in the feedstock did not have a significant impact on biogas production, despite the correct C/N ratio. In series 1, the C/N ratio ranged from 13.1 ± 0.7 in setup 6 to 30.1 ± 1.3 in setup 1, and the biogas production varied from 620.5 ± 14.6 L/kg VS in setup 1 to 923.6 ± 25.1 L/kg VS in setup 6. However in series 2, the C/N ratio achieved went from 15.1 ± 0.6 in setup 6 to 30.1 ± 1.3 in setup 1, and the biogas production increased from 620.5 ± 14.6 L/kg VS in setup 1 to 1058.8 ± 25.2 L/kg VS in setup 4.

In series 1, there was a very strong correlation between the biogas production efficiency and the C/N ratio ($r^2 = 0.8219$), (Figure 2a). However, in series 2 this relationship was less coherent ($r^2 = 0.5568$), (Figure 2a). In turn, the variation of methane production was strongly dependent on the value of the C/N ratio in series 2 ($r^2 = 0.6032$), (Figure 2b), and only moderately dependent in series 1 ($r^2 = 0.3367$), (Figure 2b).

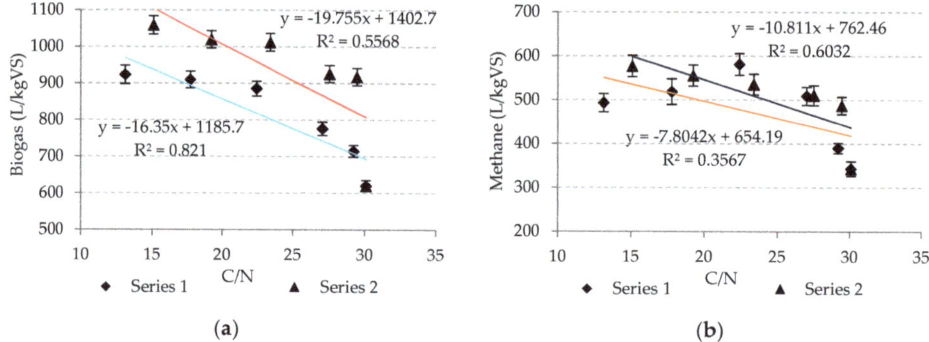


**Figure 2.** Correlation between the C/N ratio and biogas (**a**) and methane (**b**) production.

The effect of C/N ratio has been also demonstrated in studies on algae co-digestion with maize silage [46]. The highest level of biogas production (varying from 922 to 1184 mL over 30 days of anaerobic digestion) was achieved with a C/N ratio from 16 to 25. The highest content of methane in biogas of 54.9% was observed when the C/N ratio was 20, while in others setups it was about 51.0% [46].

The multiple regression models indicated that biogas production is strongly affected by the total nitrogen (TN) concentration, as well as by the amount of volatile solids (%TS) in the feedstock for anaerobic digestion. The estimated values of biogas production in the equations in relation to the results obtained in the experimental studies are very high, which indicates the correctness of the assumptions that were made, as well as the useful value of the optimization model. The regression equations for the estimation of biogas production (B) in both series of the experiment are shown in Table 7.

**Table 7.** Regression equations for the estimation of biogas production (B) with determination coefficient ($R^2$) and standard error (SE).

| Series | Formula | $R^2$ | SE |
|:---:|:---:|:---:|:---:|
| 1 | $B = 0.32TN + 114.25VS - 9459.57$ | 0.8121 | 36.965 |
| 2 | $B = 42.7TN + 397.9VS - 35416.0$ | 0.8338 | 21.871 |

$B$– biogas production (L/kgVS)
$TN$ – initial total trogen concentration in the feedstock (g/kg TS)
$VS$ – amount of VS in the feedstock (% TS)

## 4. Conclusions

It is widely claimed that the demand for renewable energy can be largely met by anaerobic digestion of biomass with different characteristics and origins. However, there are analyses that deny this claim. Unreasonable management of biomass resources may lead to an increase in greenhouse gas emissions, as well as negatively affecting the global food supply by increasing prices. Thus, there is a need to look for other sources of biomass for energy purposes that will meet the economic and ecological criteria. Microalgal biomass is an alternative to typical energy crops due to high photosynthetic efficiency of microalgae, fast rate of growth, the potential to utilize $CO_2$ emissions, resistance to various types of contamination, and the fact that microalgae can be cultured in areas that cannot be used for other purposes. In this study, the effect on anaerobic digestion performance of microalgae biomass added to feedstock mixture was analyzed.

The study showed that mixing the substrates commonly used in agricultural biogas plants (i.e., cattle slurry and maize silage) with microalgae biomass of *Arthrospira platensis* and *Platymonas subcordiformis* positively affected the final biogas production and the methane concentration in biogas.

A systematic increase was found in the biogas production with an increasing concentration of microalgae biomass ranging from 0% to 40% of VS content in the feedstock mixture for anaerobic digestion. Above this concentration, no significant increase in the biogas production was observed, and the production remained at a stable level. This was probably related to the decreasing C/N ratio in the feedstock.

It was shown that the addition of *Platymonas subcordiformis* biomass to the substrate mixture allowed us to achieve higher maximum biogas production (1058.8 ± 25.2 L/kg VS) than was obtained with *Arthrospira platensis* biomass (923.6 ± 25.1 L/kg VS). In turn, the highest methane content in biogas (over 65%) was observed in setups in which the amount of *Arthrospira platensis* biomass ranged from 20% to 40% (%VS).

There was a strong correlation between the biogas and methane production efficiencies and C/N ratio of $r^2 = 0.5568$ and $r^2 = 0.6032$ respectively, when the biomass of *Platymonas subcordiformis* was used. In turn, the relationship between biogas production and the C/N ratio was very strong ($r^2 = 0.8219$), and there was a moderate relationship between the methane production and C/N ratio ($r^2 = 0.3367$) in series with *Arthrospira platensis* biomass.

**Author Contributions:** Conceptualization, M.D. (Marcin Debowski), M.K., J.K., Z.R.-D. and M.Z.; Data curation, M.D. (Marcin Debowski), M.K., A.R. and M.D. (Magda Dudek); Formal analysis, M.D. (Marcin Debowski), M.K., J.K., A.R., M.D. (Magda Dudek), Z.R.-D. and M.Z.; Funding acquisition, M.D. (Marcin Debowski), M.D. (Magda Dudek) and M.Z.; Investigation, M.K., J.K. and M.Z.; Methodology, M.D. (Marcin Debowski), M.K., J.K., A.R. and M.D. (Magda Dudek); Project administration, M.D. (Marcin Debowski) and M.Z.; Resources, M.D. (Marcin Debowski), M.K., J.K., A.R. and Z.R.-D.; Software, M.D. (Marcin Debowski), M.K., M.D. (Magda Dudek) and Z.R.-D.; Supervision, M.D. (Marcin Debowski) and M.Z.; Validation, M.D. (Marcin Debowski), J.K., A.R., M.D. (Magda Dudek) and Z.R.-D.; Visualization, M.D. (Marcin Debowski), J.K., A.R. and Z.R.-D.; Writing-original draft, M.D. (Marcin Debowski), M.K., J.K. and A.R.; Writing-review & editing, M.D. (Marcin Debowski), M.K. and J.K. All authors have read and agreed to the published version of the manuscript.

**Funding:** The study was carried out in the framework of the project under the program BIOSTRATEG founded by the National Centre for Research and Development "Processing of waste biomass in the associated biological and chemical processes", BIOSTRATEG2/296369/5/NCBR/2016.

**Conflicts of Interest:** The authors declare no conflict of interest. The funders had no role in the design of the study; in the collection, analyses, or interpretation of data; in the writing of the manuscript, or in the decision to publish the results.

## References

1. Tursi, A. A review on biomass: Importance, chemistry, classification, and conversion. *Biofuel Res. J.* **2019**, 22, 962–979. [CrossRef]
2. Faaij, A.P.C. Securing Sustainable Resource Availability of Biomass for Energy Applications in Europe, Review of Recent Literature. Available online: http://bioenergyeurope.org/wp-content/uploads/2018/11/Bioenergy-Europe-EU-Biomass-Resources-Andr%C3%A9-Faaij-Final.pdf2018 (accessed on 15 March 2020).
3. Saratale, G.D.; Saratale, R.G.; Banu, J.R.; Chang, J.-S. Biohydrogen Production From Renewable Biomass Resources. In *Biohydrogen*, 2nd ed.; Elsevier: Amsterdam, The Netherlands, 2019; pp. 247–277.
4. Siddiqui, S.; Zerhusen, B.; Zehetmeier, M.; Effenberge, M. Distribution of specific greenhouse gas emissions from combined heat-and-power production in agricultural biogas plants. *Biomass Bioenergy* **2020**, 133, 105443. [CrossRef]
5. Daioglou, V.; Doelman, J.C.; Wicke, B.; Faaij, A. Integrated assessment of biomass supply and demand in climate change mitigation scenarios. *Glob. Environ. Chang.* **2019**, 54, 88–101. [CrossRef]

6.  Johanson, D.; Azar, C.A. Scenario based analysis of land competition between food and bioenergy production in the US. *Clim. Chang.* **2007**, *82*, 267–291. [CrossRef]

7.  Goyal, H.B.; Seal, D.; Saxena, R.C. Bio-fuels from thermochemical conversion of renewable resources: A review. *Renew. Sustain. Energy Rev.* **2008**, *12*, 504–517. [CrossRef]

8.  Schenk, P.M.; Thomas-Hall, S.R.; Stephens, E.; Marx, U.C.; Mussgnug, J.H.; Posten, C.; Kruse, O.; Hankamer, B. Second generation biofuels: High efficiency microalgae for biodiesel production. *Bioenergy Res.* **2008**, *1*, 20–43. [CrossRef]

9.  Wirth, R.; Lakatos, G.; Maróti, G.; Bagi, Z.; Minárovics, J.; Nagy, K.; Kondorosi, E.; Rákhely, G.; Kovács, K.L. Exploitation of algal-bacterial associations in a two-stage biohydrogen and biogas generation process. *Biotechnol. Biofuels* **2015**, *8*, 59. [CrossRef]

10. Muhammad, G.; Alam, M.A.; Xiong, W.; Lv, Y.; Xu, J.L. Microalgae Biomass Production: An Overview of Dynamic Operational Methods. In *Microalgae Biotechnology for Food, Health and High Value Products*; Springer: Singapore, 2020; pp. 415–432.

11. Yen, H.-W.; Brune, D.E. Anaerobic co-digestion of algal sludge and waste paper to produce methane. *Bioresour. Technol.* **2007**, *98*, 130–134. [CrossRef]

12. Samson, R.; Leduy, A. Biogas production from anaerobic digestion of Spirulina maxima algal biomass. *Biotechnol. Bioeng.* **1982**, *24*, 1919–1924. [CrossRef]

13. Samson, R.; Leduy, A. Detailed study of anaerobic digestion of Spirulina maxima algal biomass. *Biotechnol. Bioeng.* **1986**, *28*, 1014–1023. [CrossRef]

14. Ras, M.; Lardon, L.; Bruno, S.; Bernet, N.; Steyer, J.P. Experimental study on a coupled process of production and anaerobic digestion of Chlorella vulgaris. *Bioresour. Technol.* **2011**, *102*, 200–206. [CrossRef] [PubMed]

15. Uziel, M. Solar Energy Fixation and Conversion with Algal Bacterial Systems. Ph.D. Thesis, University of California, Berkeley, CA, USA, 1978.

16. Chynoweth, D.P.; Turick, C.E.; Owens, J.M.; Jerger, D.E.; Peck, M.W. Biochemical methane potential of biomass and waste feedstocks. *Biomass Bioenergy* **1993**, *5*, 95–111. [CrossRef]

17. Wise, D.L.; Augenstein, D.C.; Ryther, J.H. Methane fermentation of aquatic biomass. *Resour. Recovery Conserv.* **1979**, *4*, 217–237. [CrossRef]

18. Bruhn, A.; Dahl, J.; Nielsen, H.B.; Nikolaisen, L.; Rasmussen, M.B.; Markager, S.; Olesen, B.; Arias, C.; Jensen, P.D. Bioenergy potential of Ulvalactuca: Biomass yield, methane production and combustion. *Bioresour. Technol.* **2011**, *102*, 2595–2604. [CrossRef]

19. Legros, A.; Marzano, C.M.A.D.; Naveau, H.P.; Nyns, E.J. Fermentation profiles in bioconversions. *Biotechnol. Lett.* **1983**, *5*, 7–12. [CrossRef]

20. Hernandez, E.P.S.; Cordoba, L.T. Anaerobic digestion of chlorella vulgaris for energy production. *Resour. Conserv. Recycl.* **1993**, *9*, 127–132. [CrossRef]

21. Vergara-Ferna'ndez, A.; Vargas, G.; Alarco'n, N.; Antonio, A. Evaluation of marine algae as a source of biogas in a two-stage anaerobic reactor system. *Biomass Bioenergy* **2008**, *32*, 338–344. [CrossRef]

22. Rittmann, B.E. Opportunities for renewable bioenergy using microorganisms. *Biotechnol. Bioeng.* **2008**, *100*, 203–212. [CrossRef]

23. Stephens, E.; Ross, I.L.; King, Z.; Mussgnug, J.H.; Kruse, O.; Posten, C.; Borowitzka, M.A.; Hankamer, B. An economic and technical evaluation of microalgal biofuels. *Nat. Biotechnol.* **2010**, *28*, 126–128. [CrossRef]

24. Harun, R.; Davidson, M.; Doyle, M.; Gopiraj, R.; Danquah, M.; Forde, G. Technoeconomic analysis of an integrated microalgae photobioreactor, biodiesel and biogas production facility. *Biomass Bioenergy* **2011**, *35*, 741–747. [CrossRef]

25. Chen, W.-H.; Lin, B.-J.; Huang, M.-Y.; Chang, J.-S. Thermochemical conversion of microalgal biomass into biofuels: A review. *Bioresour. Technol.* **2015**, *184*, 314–327. [CrossRef] [PubMed]

26. Böjti, T.; Kovács, K.L.; Kakuk, B.; Wirth, R.; Rákhely, G.; Bagi, Z. Pretreatment of poultry manure for efficient biogas production as monosubstrate or co-fermentation with maize silage and corn stover. *Anaerobe* **2017**, *46*, 138–145. [CrossRef] [PubMed]

27. Mata-Alvarez, J.; Macé, S.; Llabrés, P. Anaerobic digestion of organic solid wastes. An overview of research achievements and perspectives. *Bioresour. Technol.* **2000**, *74*, 3–16. [CrossRef]

28. Wu, X.; Yao, W.; Zhu, J.; Miller, C. Biogas and CH4 productivity by co-digesting swine manure with three crop residues as an external carbon source. *Bioresour. Technol.* **2010**, *101*, 4042–4047. [CrossRef]

29. Carver, S.M.; Hulatt, C.J.; Thomas, D.N.; Tuovinen, O.H. Thermophilic, anaerobic co-digestion of microalgal biomass and cellulose for H2 production. *Biodegradation* **2011**, *22*, 805–814. [CrossRef]

30. González-Delgado, A.D.; Kafarov, V. Microalgae based biorefinery: Issues to consider. A review. *CT&F-Ciencia Tecnología y Futuro* **2011**, *4*, 5–22.

31. Giuliano, A.; Bolzonella, D.; Pavan, P.; Cavinato, C.; Cecchi, F. Co-digestion of livestock effluents, energy crops and agro-waste: Feeding and process optimization in mesophilic and thermophilic conditions. *Bioresour. Technol.* **2013**, *128*, 612–618. [CrossRef]

32. Amon, T.; Amon, B.; Kryvoruchko, V.; Zollitsch, W.; Mayer, K.; Gruber, L. Biogas production from maize and dairy cattle manure—Influence of biomass composition on the methane yield. *Agric. Ecosyst. Environ.* **2007**, *118*, 173–182. [CrossRef]

33. Kalamaras, S.D.; Kotsopoulos, T.A. Anaerobic co-digestion of cattle manure and alternative crops for the substitution of maize in South Europe. *Bioresour. Technol.* **2014**, *172*, 68–75. [CrossRef]

34. Matsui, T.; Koike, Y. Methane fermentation of a mixture of seaweed and milk at a pilot-scale plant. *J. Biosci. Bioeng.* **2010**, *110*, 558–563. [CrossRef]

35. Mussgnug, J.H.; Klassen, V.; Schlüter, A.; Kruse, O. Microalgae as substrates for fermentative biogas production in a combinedbiorefinery concept. *J. Biotechnol.* **2010**, *150*, 51–56. [CrossRef] [PubMed]

36. Singh, J.; Gu, S. Commercialization potential of microalgae for biofuels production. *Renew. Sustain. Energy Rev.* **2010**, *14*, 2596–2610. [CrossRef]

37. Parmar, A.; Singh, N.K.; Pandey, A.; Gnansounou, E.; Madamwar, D. Cyanobacteria and microalgae: A positive prospect for biofuels. *Bioresour. Technol.* **2011**, *102*, 10163–10172. [CrossRef] [PubMed]

38. Elser, J.J.; Fagan, W.F.; Denno, R.F.; Dobberfuhl, D.R.; Folarin, A.; Huberty, A.; Interlandi, S.; Kilham, S.S.; McCauley, E.; Schulz, K.L.; et al. Nutritional constraints in terrestrial and freshwater food webs. *Nature* **2000**, *408*, 578–580. [CrossRef]

39. Corno, L.; Pilu, R.; Tambone, F.; Scaglia, B.; Adani, F. New energy crop giant cane (*Arundo donax* L.) can substitute traditional energy crops increasing biogas yield and reducing costs. *Bioresour. Technol.* **2015**, *191*, 197–204. [CrossRef]

40. Schwede, S.; Kowalczyk, A.; Gerber, M.; Span, R. Anaerobic co-digestion of the marine microalga *Nannochloropsis salina* with energy crops. *Bioresour. Technol.* **2013**, *148*, 428–435. [CrossRef]

41. Parkin, G.F.; Owen, W.F. Fundamentals of anaerobic digestion of wastewater sludges. *J. Environ. Eng.* **1986**, *112*, 867–920. [CrossRef]

42. Deublein, D.; Steinhauser, A. *Biogas from Waste and Renewable Resources*; WILEY-VCH Verlag GmbH & Co. KGaA: Weinheim, Germany, 2008.

43. Burton, C.; Turner, C. *Manure Management*; Silsoe Research Institute: Silsoe, Bedfordshire, UK, 2003; pp. 281–282.

44. Mshandete, A.; Kivaisi, A.; Rubindamayugi, M.; Mattiasson, B. Anaerobic batch co-digestion of sisal pulp and fish wastes. *Bioresour. Technol.* **2004**, *95*, 19–24. [CrossRef]

45. Abbasi, T.; Tauseef, S.M.; Abbasi, S.A. *Biogas Energy*; Springer: New York, NY, USA, 2012.

46. Zhong, W.; Zhongzhi, Z.; Yijing, L.; Wei, Q.; Meng, X.; Min, Z. Biogas productivity by co-digesting Taihu blue algae with corn straw as an external carbon source State Key. *Bioresour. Technol.* **2012**, *114*, 281–286. [CrossRef]

*Article*

# Simulation of the Growth Potential of Sugarcane as an Energy Crop Based on the APSIM Model


**Ting Peng** [1,2,†], **Jingying Fu** [1,2,†], **Dong Jiang** [1,2,3,*] and **Jinshuang Du** [1,2,*]

1. Institute of Geographical Sciences and Natural Resources Research, Chinese Academy of Sciences, Beijing 100101, China; pengt.19s@igsnrr.ac.cn (T.P.); fujy@igsnrr.ac.cn (J.F.)
2. College of Resources and Environment, University of Chinese Academy of Sciences, Beijing 100049, China
3. Key Laboratory of Carrying Capacity Assessment for Resource and Environment, Ministry of Land & Resources, Beijing 100101, China
* Correspondence: jiangd@igsnrr.ac.cn (D.J.); jins.18s@igsnrr.ac.cn (J.D.)
† These authors contributed equally to this work.


Received: 18 March 2020; Accepted: 27 April 2020; Published: 1 May 2020


**Abstract:** Research on the development of plants grown for energy purposes is important for ensuring the global energy supply and reducing greenhouse gas emissions, and simulation is an important method to study its potential. This paper evaluated the marginal land that could be used to grow sugarcane in the Guangxi Zhuang Autonomous Region. Based on the meteorological data from 2009 to 2017 in this region and field observations from sugarcane plantations, the sensitivity of the APSIM (Agricultural Production Systems sIMulator) model parameters was analyzed by an extended Fourier amplitude sensitivity test, and the APSIM model was validated for sugarcane phenology and yields. During the process of model validation, the value of the determination coefficient $R^2$ of the observed and simulated values was between 0.76 and 0.91, and the consistency index $D$ was between 0.91 and 0.97, indicating a good fit. On this basis, the APSIM sugarcane model was used to simulate the sugarcane production potential of the marginal land on a surface scale, and the distribution pattern of sugarcane production potential in the marginal land was obtained. The simulation results showed that if sugarcane was planted as an energy crop on the marginal land in Guangxi, it would likely yield approximately $42,522.05 \times 10^4$ t of cane stalks per year. It was estimated that the sugarcane grown on the marginal land plus 50% of the sugarcane grown on the cropland would be sufficient to produce approximately $3847.37 \times 10^4$ t of ethanol fuel. After meeting the demands for vehicle ethanol fuel in Guangxi, $3808.14 \times 10^4$ t of ethanol fuel would remain and could be exported to the ASEAN (Association of Southeast Asian Nations).

**Keywords:** APSIM sugarcane model; energy potential; marginal land; sensitivity analysis


---

## 1. Introduction

Since 2011, China has been the largest energy consumer in the world. Due to the rapid growth of its population and GDP (gross domestic product), the foreign dependence rates for oil and natural gas are approximately 61% and 33%, respectively [1]. The long-term exploitation and utilization of fossil energy in China, especially coal and oil, has caused a large amount of greenhouse gas emissions, which will inevitably impact the ecological environment in China and that of the rest of the world [2]. Therefore, China will face two major problems in the future, namely, fossil energy shortages and environmental pollution. To meet energy needs and ensure sustainable development, China is in urgent need of bioliquid fuels, including ethanol liquid fuel [3,4]. However, due to the Sino–US trade war, China has not imported US ethanol fuel since July 2018. The ethanol fuel imports of China are severely constrained, and the demand is high. Therefore, the bioliquid ethanol fuel industry is still the focus of future biomass energy development in China. Since 2007, China has issued a series of policies emphasizing the developmental needs and plans for ethanol fuel from non-grain plants [5,6]. In this context, Guangxi vigorously promotes the industrialization and development of non-grain ethanol fuel plants, such as cassava and sugarcane. Due to its subtropical climate conditions and rich biological resources, Guangxi has been at the forefront of the non-grain biomass energy industry in China. In November 2019, Zhanjiang Customs of Guangdong Province reported that 3062.17 t of ethanol fuel was exported by a bioenergy company to Vietnam and was issued a China-ASEAN (Association of Southeast Asian Nations) Certificate of Origin. China signed the ASEAN Free Trade Agreement in 2002 [7], and this was the first zero-tariff export of Chinese ethanol fuel producers to the international market. As a bridgehead to the ASEAN, the non-grain ethanol fuel industry in Guangxi will be further developed through border trade.

Although the development and utilization of biomass energy in China has started late, its development has been relatively fast. Research on distribution, selection, cultivation, improvement, and processing technology and equipment for biomass energy crops has made great achievements which will benefit the further development of biomass energy in China [8–10]. Sugarcane is a high biomass crop, which has many advantages as a biomass energy crop, such as a high yield per unit area, high light energy storage efficiency, and a relatively low processing cost. Therefore, it is necessary to study the production potential and energy efficiency of sugarcane.

Few studies have evaluated the potential bioenergy that can be produced from sugarcane. At present, there are two main methods of sugarcane yield prediction, namely, by remote sensing data combined with a geographic information system (GIS) and secondly by model evaluation. Cervi et al. used a spectral ratio of remote sensing data for vegetation assessment, i.e., the normalized difference vegetation index, in order to estimate the spatial yield of sugarcane [11]. Singels et al. used a land surface energy balance algorithm (SEBAL) to estimate the biomass yield of sugarcane from remote sensing data [12]. Yawson et al. used satellite remote sensing data and geographic information systems to assess sugarcane yield supply potential [13]. These studies only focused on the spatial distribution of potential bioenergy and failed to study the potential effects of land use and climate change. Lisboa et al. developed a prediction model of sugarcane yield based on a normalized difference vegetation index (NDVI) and leaf tissue nutrient concentration data to help the Brazilian sugarcane sector monitor crop yield changes [14]. Based on the statistical analysis of the data, Satiro et al. established a model to predict sugarcane yield from soil properties [15]. Dias et al. used three different sugarcane simulation models (FAO (Food and Agriculture Organization) agroecological Zone, DSSAT (Decision Support System for Agrotechnology Transfer)/CANEGRO and APSIM (Agricultural Production Systems sIMulator) sugarcane) to estimate potential and water-restricted yields and production gaps for 30 locations in Brazil [16]. These studies provide a point-scale accurate estimation of sugarcane yield by establishing mathematical models, but it is difficult to achieve surface-scale simulations. Other researchers have considered land and water properties in order to assess sugarcane growth potential. Rodriguez et al. assessed the potential for biofuel crop expansion by combining the water footprint, water availability, and land availability [17]. Sanches et al. used soil attributes to estimate sugarcane yield, and their research showed that the soil ECa (apparent electrical conductivity) in sugarcane fields when mapped by electromagnetic induction sensors could reflect the potential yield well [18]. However, these studies have failed to consider the sustainable development of bioenergy. The main objective of this paper is to simulate and evaluate the production potential of the sugarcane as an energy crop on the marginal land of the Guangxi Zhuang Autonomous Region by combining an APSIM sugarcane model and GIS spatial analysis technology.

## 2. Methods and Data

In this study, eight years of meteorological, soil, and field observation data were collected for the APSIM model parameter sensitivity analysis, parameter calibration, and verification and surface scale simulation. See Section 2.2 for details concerning the data. On this basis, the paper first extracts the marginal land suitable for sugarcane cultivation in Guangxi, and then performs parameter sensitivity analysis, parameter calibration, and model validity verification of the APSIM sugarcane model. Then, the APSIM sugarcane model is used to simulate sugarcane production on marginal land on a surface scale. Finally, the distribution of sugarcane production potential on the marginal land in Guangxi is obtained, and the sustainable development of bioenergy from sugarcane is analyzed. The analysis framework for the potential production of bioenergy from sugarcane is shown in Figure 1.

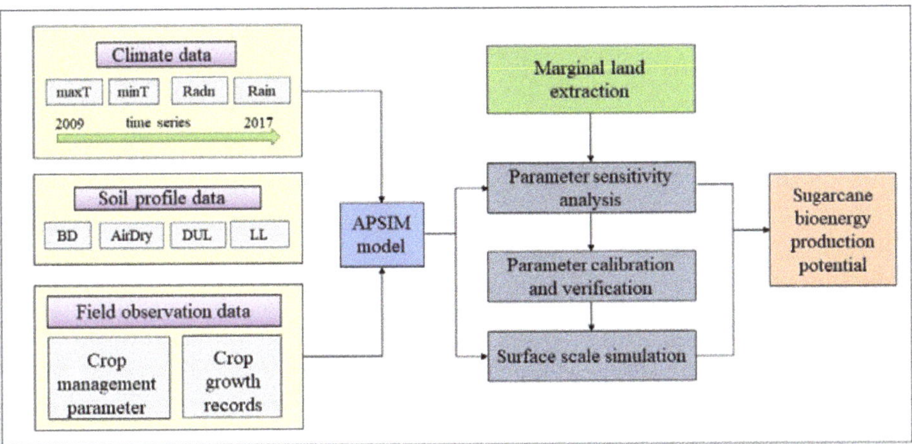


**Figure 1.** Analysis framework used in the current study for the potential production of bioenergy from sugarcane. The meanings of the abbreviations maxT, minT, Radn, and Rain refer to the maximum temperature, minimum temperature, solar radiation, and rainfall, respectively. The abbreviations BD, AirDry, DUL, and LL refer to the bulk density, air-dried soil moisture content, field capacity and wilting coefficient, respectively.

*2.1. Study Sites*

As the region with the highest proportion of sugarcane planting area in China [19], the Guangxi Zhuang Autonomous Region was chosen here to study the growth potential of sugarcane as an energy crop with the APSIM model. The study area is located in Southern China (see Figure 2) and exists at a low altitude and has a subtropical monsoon climate [20]. The area of cultivated land in Guangxi is about 59,724 km$^2$, of which 8864 km$^2$ is planted with sugarcane, and the area of unused land (mainly including shrub forest land, sparse forest land, and grassland) is about 85,938 km$^2$. The introduction of the extraction of marginal land suitable for sugarcane cultivation from unused land is located in Section 2.2.1. In the following research, the existing cultivated and unused land in Guangxi will be combined to analyze the potential of sugarcane as an energy crop.

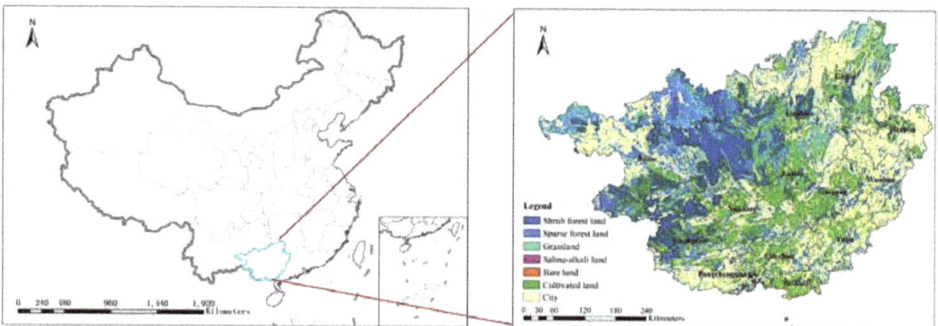

**Figure 2.** Geographical location and land cover distribution of the Guangxi Zhuang Autonomous Region.

*2.2. Methodology and Data Sources*

The data used in this study are the marginal land suitable for sugarcane growth and the inputs for running the APSIM model, including information on the soil, weather, crop variety, and field management.

2.2.1. Marginal Land Data

The Department of Science and Technology of the Ministry of Agriculture of China has defined the marginal land for energy crops as winter fallow fields and wastelands. Suitable wasteland refers to open forest land, natural grassland, shrublands, and unused land that is suitable for cultivating energy crops [21].

The quantity and spatial distribution of land resources that are suitable for growing non-grain ethanol fuel raw sugarcane in China were extracted by a multi-factor comprehensive evaluation method and socioeconomic factor limitation method that considers the growth characteristics of sugarcane. The specific technical methods are as follows [22]:

(1) According to the principle of avoiding biofuel development that competes with people for food and that which competes with grain for land, as issued by the Ministry of Agriculture in 2007, arable land was excluded.

(2) To protect the ecological environment and prevent the destruction of ecosystems, land types such as nature reserves, landscape, historical sites, and protected zones were excluded.

(3) Taking into account the development needs for animal husbandry in China, the high- and medium-coverage grasslands of the five pasture areas in China were excluded.

(4) According to the characteristics of land resources suitable for energy crop development, land use types such as swamp land, water bodies, and construction land were excluded. Land use types suitable for the cultivation of energy crops include shrub forest land, sparse forest land, grassland, mudflats, saline-alkali land, and bare land.

(5) Based on the relevant literature [23], the growth characteristics of sugarcane were analyzed (see Table 1) and an index system of the natural conditions for sugarcane growth was established. Setting the lower limit of sugarcane requirements for soil, temperature, moisture, slope, and other conditions, the GIS technology was used to extract the land resources that were suitable for sugarcane planting.

**Table 1.** Sugarcane growth conditions based on the relevant literature [23].

| Factor | | Unit | Suitable Conditions |
|---|---|---|---|
| Slope | | | <25 |
| Soil organic matter content | | % | ≥2 |
| Soil type | | - | Loam or sandy loam |
| pH | | - | ≈4.5–8.0 |
| Duration of sunshine | | h | ≥1195 |
| Cumulative temperature in base 10 °C | | °C | ≈6500–8000 |
| Jointing stage to maturity | | - | ≈18–25 |
| Average annual air humidity | | % | Approximately 60% |
| Annual precipitation | | mm | ≈800–1200 |
| Temperature | Germination | °C | ≥13 |
| | Germination to seedling | °C | ≈20–25 |
| | Seedling stage to leaf stage | °C | ≈20–30 |
| | Leaf stage to jointing stage | °C | ≈25–28 |

Referring to the above steps, the marginal land resources suitable for planting sugarcane in China were extracted (see Figure 3). The marginal land resources suitable for planting sugarcane in the Guangxi Zhuang Autonomous Region were obtained by the cutting operation of the ArcGIS (https://developers.arcgis.com/) software, as shown in Figure 4.

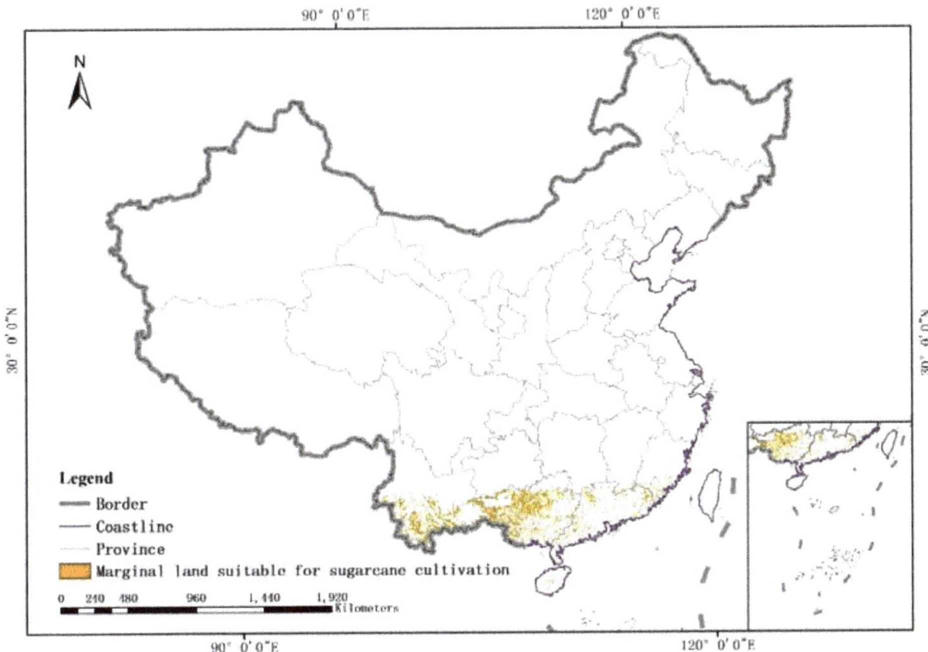


**Figure 3.** Spatial distribution of marginal land suitable for growing sugarcane in China.

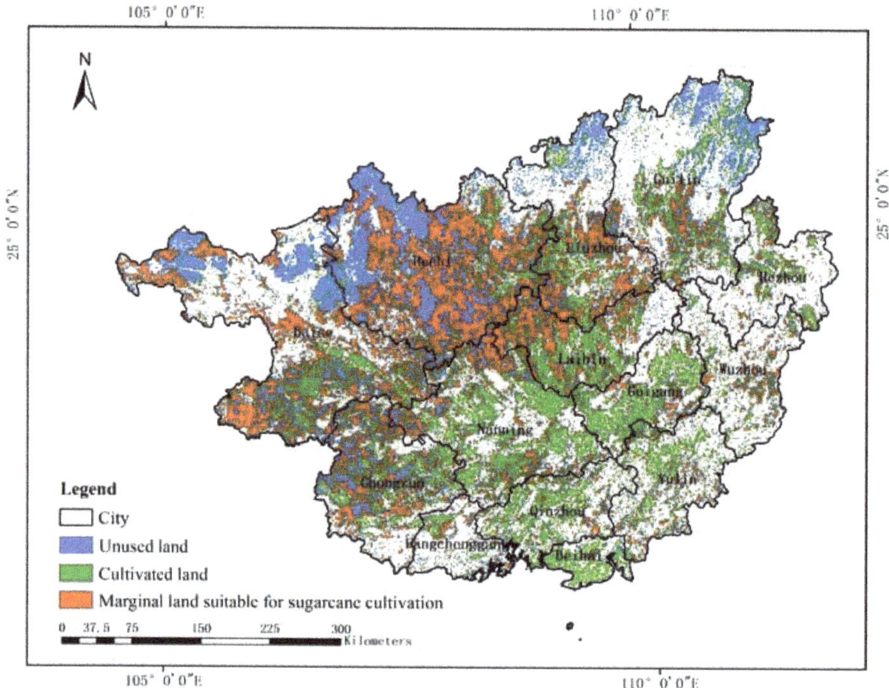


**Figure 4.** Spatial distribution of marginal land suitable for growing sugarcane, cultivated land, and remaining unused land in the Guangxi Zhuang Autonomous Region.

### 2.2.2. Field Observation Data

This study uses field observation data from Ruan [24] to calibrate the crop variety parameters. The sugarcane variety tested in the field was Guitang 32 (Guitang 02-208) cultivated by the Sugarcane Research Institute of the Guangxi Academy of Agricultural Sciences. The female parent of this variety is Yuetang 91-976 and the male parent is Xintaitang 1. Guitang 32 is an early-maturing sugarcane cultivar. It grows vigorously and needs sufficient basic fertilizer for cultivation. Its suitable planting time is from late February to mid-March [25].

The field test period was from 2015 to 2017, and there were four test plots. For each plot, the planting area was 6 m wide and 20 m long, and the plot area was 120 m$^2$. Each plot was planted with five rows, the row spacing was 1.2 m, and the planting density was 105,000 buds·ha$^{-1}$. The field management measures for each plot were the same and 300 kg·ha$^{-1}$ of nitrogen fertilizer was used during the whole growth period. The base and topdressing fertilizers were applied at a 1:1 ratio, while the potash (330 kg·ha$^{-1}$) and phosphate (150 kg·ha$^{-1}$) fertilizers were applied once, with the base fertilizer. The other field management measures were the same as those used in conventional agricultural production, and there was no irrigation during the growth period.

Field observations included the crop growth period, the leaf area index of each growth stage, the dry weight of the aboveground part, the yield of the sugarcane stem, etc. The sugarcane growth records are shown in Table 2.

After the completion of the model crop variety parameter calibration, the parameters of the APSIM sugarcane model needed to be verified, so more sugarcane field observation records were needed. In this study, the sugarcane field observation data from Zu et al. [26] were used to verify the results of the model parameter calibration, mainly to verify the sugarcane stalk yield and the phenological period of the sugarcane. Model calibration and validation data can be seen in Table 3.

**Table 2.** Growth parameters measured at the sugarcane field experiment from 2015 to 2017.

| Plant Type | Sowing Date | Emergence Date | Cane Appearance Date | Harvest Date | Leaf Area Index | Dry Weight (t·ha$^{-1}$) | Cane Stem Yield (t·ha$^{-1}$) | Sugar Dry Weight (t·ha$^{-1}$) |
|---|---|---|---|---|---|---|---|---|
| Plant | 2015-02-03 | 2015-04-27 | 2015-07-24 | 2015-12-28 | 6.15 | 50.43 | 119.65 | 10.32 |
| Ratoon | 2015-12-28 | 2016-02-22 | 2016-06-29 | 2017-01-06 | 5.45 | 45.75 | 118.53 | 9.24 |
| Plant | 2015-03-15 | 2015-05-23 | 2015-08-08 | 2016-01-17 | 5.96 | 42.84 | 110.76 | 8.98 |
| Plant | 2015-04-16 | 2015-06-27 | 2015-07-08 | 2016-02-22 | 4.77 | 49.56 | 115.77 | 9.11 |
| Plant | 2016-02-28 | 2016-05-06 | 2016-07-30 | 2017-01-06 | 5.24 | 52.18 | 116.34 | 10.03 |

**Table 3.** Model calibration and validation data [24,26].

| | Site | Plant Year | Plant Type | Sugarcane Stalk Yield (t·ha$^{-1}$) | Phenological Period | | | |
|---|---|---|---|---|---|---|---|---|
| | | | | | Sowing Date | Emergence Date | Cane Appearance Date | Harvest Date |
| Calibration | Nanning | 2015 | Plant | 119.65 | 2015-02-03 | 2015-04-26 | 2015-07-24 | 2015-12-28 |
| | Nanning | 2015/2016 | Ratoon | 118.53 | 2015-12-29 | 2016-02-22 | 2016-06-29 | 2017-01-06 |
| | Laibin | 2009/2010 | Ratoon | 62.83 | 2009-11-11 | 2010-03-03 | 2010-06-14 | 2010-11-10 |
| | Shatang | 2010 | Ratoon | 77.23 | 2010-02-26 | 2010-05-10 | 2010-07-20 | 2010-12-20 |
| | Guigang | 2010/2011 | Ratoon | 66.21 | 2010-11-21 | 2011-04-14 | 2011-06-30 | 2011-11-30 |
| | Yizhou | 2011/2012 | Ratoon | 54.43 | 2011-12-03 | 2012-04-16 | 2012-06-06 | 2012-11-10 |
| | Pingguo | 2011/2012 | Ratoon | 68.77 | 2011-12-11 | 2012-03-28 | 2012-06-14 | 2012-11-10 |
| Validation | Laibin | 2010/2011 | Ratoon | 61.76 | 2010-11-11 | 2011-04-18 | 2011-06-22 | 2011-11-10 |
| | Laibin | 2011/2012 | Ratoon | 67.71 | 2011-11-11 | 2012-04-10 | 2012-06-06 | 2012-11-07 |
| | Pingguo | 2009/2010 | Ratoon | 49.97 | 2009-11-11 | 2010-03-17 | 2010-06-23 | 2010-11-20 |
| | Pingguo | 2011 | Plant | 39.47 | 2011-02-27 | 2011-04-17 | 2011-06-30 | 2011-12-10 |
| | Shatang | 2011/2012 | Ratoon | 82.50 | 2011-12-21 | 2012-04-16 | 2012-06-14 | 2012-11-20 |
| | Shatang | 2012/2013 | Ratoon | 83.00 | 2012-11-21 | 2013-03-18 | 2013-06-14 | 2013-11-20 |
| | Guigang | 2009/2010 | Ratoon | 62.79 | 2009-11-20 | 2010-04-03 | 2010-06-18 | 2010-11-20 |
| | Guigang | 2011/2012 | Ratoon | 71.74 | 2011-12-01 | 2012-03-30 | 2012-06-06 | 2012-11-20 |
| | Yizhou | 2009/2010 | Ratoon | 60.25 | 2009-10-21 | 2010-03-10 | 2010-06-20 | 2010-11-10 |
| | Yizhou | 2011 | Plant | 53.78 | 2011-02-23 | 2011-04-14 | 2011-07-02 | 2011-12-02 |
| | Nanning | 2015 | Plant | 110.76 | 2015-03-15 | 2015-05-23 | 2015-08-08 | 2016-01-17 |
| | Nanning | 2016 | Plant | 116.34 | 2016-02-28 | 2016-05-06 | 2016-07-30 | 2017-01-26 |

### 2.2.3. Meteorological Data

The APSIM sugarcane model requires daily meteorological data as an input. The time span needs to cover the entire growth period of the crop. The meteorological file in the APSIM model contains 10 data items, including the site name, latitude, and eight meteorological data items (see Table 4). The daily solar radiation, maximum temperature, minimum temperature, and precipitation were obtained by interpolating the meteorological station data through the ANUSPLIN version 4.3 software [27]. The data from 2009 to 2017 were collected from the National Meteorological Science Data Sharing Service Platform (http://data.cma.cn/).

**Table 4.** Meteorological file data items of the APSIM (Agricultural Production Systems sIMulator) sugarcane model.

| Name | Definition | Unit | Remarks |
|---|---|---|---|
| Site | - | - | |
| Latitude | - | Decimal | |
| Tav | Annual average temperature | °C | |
| Amp | Monthly average temperature annual amplitude | °C | |
| Year | - | - | |
| Day | - | - | |
| Radn | Solar radiation | $(MJ \cdot m^{-2})$ | Non-negative |
| Maxt | Maximum temperature | (°C) | |
| Mint | Minimum temperature | (°C) | |
| Rain | Rainfall | (mm) | Non-negative |

### 2.2.4. Soil Profile Data

The soil profile data used in this paper mainly include soil hydrological properties and soil nitrogen. Soil hydrological properties include the saturated water content, field capacity, permanent wilting coefficient, and air-dried soil moisture content of each soil profile. Soil nitrogen properties include the nitrate nitrogen content, ammonia nitrogen content, pH value, and organic carbon content (see Table 5). In the process of model calibration and verification, soil parameters were derived from field observation data [24]. In the process of model surface scale simulation, soil hydraulic parameters were obtained from the database of soil hydraulic parameters established by Dai et al. [28], and soil nitrogen parameters used the soil nitrogen values of field observation data. The soil profile attribute data are divided into seven layers at depths of 4.5, 9.1, 16.6, 28.9, 49.3, 82.9, and 138.3 cm.

**Table 5.** Soil characteristic parameters of the APSIM sugarcane model.

| Name | Description | Unit |
|---|---|---|
| Depth | Layer depth | cm |
| AirDry | Air-dried soil moisture content | $cm^3 \cdot cm^{-3}$ |
| BD | Bulk density | $g \cdot cm^{-3}$ |
| LL15 | Permanent wilting coefficient | $cm^3 \cdot cm^{-3}$ |
| DUL | Field capacity | $cm^3 \cdot cm^{-3}$ |
| SAT | Saturated water content | $cm^3 \cdot cm^{-3}$ |
| $NO_3$ | Nitrate nitrogen content | ppm |
| $NH_4$ | Ammonia nitrogen content | ppm |
| pH | Potential of hydrogen | - |
| OC | Organic carbon content | % |

### 2.3. APSIM Module

#### 2.3.1. APSIM and Sugarcane Module

The APSIM model is a comprehensive mechanistic model that was developed by CSIRO (Commonwealth Scientific and Industrial Research Organisation) and APSRU (Agricultural Production System Research Unit) in 1991 to simulate the biophysical processes of agricultural production systems. The APSIM sugarcane model is a built-in sugarcane module in the APSIM model. It can interact with the soil, agricultural residue, and agricultural management modules to automatically simulate the water, fertilizer, and nutrient cycling between soil and sugarcane crops.

#### 2.3.2. Model Localization Settings

The APSIM model used in this paper was version 7.10. The input parameters include three aspects, namely, the meteorological, soil, and crop variety parameters.

Meteorological Data

The meteorological data used in the study were obtained from the National Meteorological Information Center and the APSIM meteorological data files were processed in R and Python.

Soil Data

The soil parameters for the process of parameter adjustment and model validation were derived from the data collected from the experimental sites [24], as shown in Table 6. See Table 5 for the description of the parameters in Table 6.

**Table 6.** Soil properties observed in the field experiment.

| Depth (cm) | BD ($g \cdot cm^{-3}$) | AirDry ($cm^3 \cdot cm^{-3}$) | DUL ($cm^3 \cdot cm^{-3}$) | LL15 ($cm^3 \cdot cm^{-3}$) | SAT ($cm^3 \cdot cm^{-3}$) | OC (%) | pH | $NH_4^+$-N (ppm) | $NO_3^-$-N (ppm) |
|---|---|---|---|---|---|---|---|---|---|
| 0–10 | 1.51 | 0.08 | 0.32 | 0.11 | 0.36 | 1.57 | 5.56 | 75.16 | 87.59 |
| 10–20 | 1.52 | 0.08 | 0.31 | 0.11 | 0.35 | 2.28 | 5.65 | 122.0 | 62.45 |
| 20–30 | 1.54 | 0.09 | 0.30 | 0.12 | 0.34 | 2.75 | 6.33 | 59.65 | 86.03 |
| 30–45 | 1.63 | 0.10 | 0.28 | 0.13 | 0.31 | 2.88 | 5.81 | 72.16 | 97.77 |
| 45–60 | 1.59 | 0.10 | 0.28 | 0.15 | 0.32 | 2.12 | 5.78 | 103.64 | 45.92 |
| 60–90 | 1.55 | 0.14 | 0.28 | 0.15 | 0.32 | 1.66 | 5.71 | 128.37 | 55.99 |
| 90–130 | 1.55 | 0.14 | 0.27 | 0.15 | 0.31 | 1.46 | 6.01 | 132.43 | 44.87 |

Crop Variety Data

This study used the Australian sugarcane variety Q138 because this variety is suitable for planting in high temperature and rainy areas, featuring good vitality and strong adaptability. The crop variety parameters of the APSIM sugarcane model and the default values of Q138 are shown in Table 7.

### 2.4. Simulation of Sugarcane Production Process

The APSIM model was spatially expanded by GIS technology to achieve surface-scale simulation of sugarcane on the marginal land in Guangxi. According to the scheme in Section 2.2.1, the surface vector data (polygon shapefile format) of the marginal land suitable for sugarcane planting were obtained in the ArcGIS (version 10.3.1) software. Since the APSIM model can only perform point simulation, it is necessary to convert the surface into points. Therefore, the "raster extracted by mask" and "raster resample" GIS technologies were used to rasterize the marginal land (into raster format). Python (version 3.7.4) and R (version 3.6.1) [29] were used to process the meteorological and soil data and connect these attributes to the grid point of the marginal land. Program modules such as "etree", "pandas", and "data.table" were utilized in this process. Then, "raster to point" GIS technology was used to transform the raster data into point vector data (point shapefile format). As a consequence, the surface simulation of the APSIM model in the study area was transformed into tens of thousands of point simulations. The Python programming language was used for data formatting, batch operation, and model output sorting during the model operation. Finally, the "vector to raster", "hierarchical rendering", and "thematic map drawing" GIS technologies were used to visualize the running results of the model. The technical process of the model space extension is shown in Figure 5.

**Table 7.** Default values of sugarcane variety parameters for the Q138 sugarcane variety in the APSIM model.

| Name | Description | Unit | Default Value |
|---|---|---|---|
| cane_fraction | Percentage of daily biomass allocated to cane stems | % | 0.70 |
| sucrose_delay | Sugarcane sugar accumulation delay factor | g·m$^{-2}$ | 600.0 |
| min_sstem_sucrose | Minimum cane stalk biomass before sugar accumulation begins | g·m$^{-2}$ | 1500 |
| min_sstem_sucrose_redn | Minimum sugarcane stem accumulation reduced under stress | g·m$^{-2}$ | 10 |
| tt_emerg_to_begcane | Accumulated temperature from emergence to jointing | °C·d | 1900 |
| tt_begcane_to_flowering | Accumulated temperature from jointing to flowering | °C·d | 6000 |
| tt_flowering_to_crop_end | Accumulated temperature from flowering to maturity | °C·d | 2000 |
| green_leaf_no | The maximum number of green leaves before plant maturation | - | 13.0 |

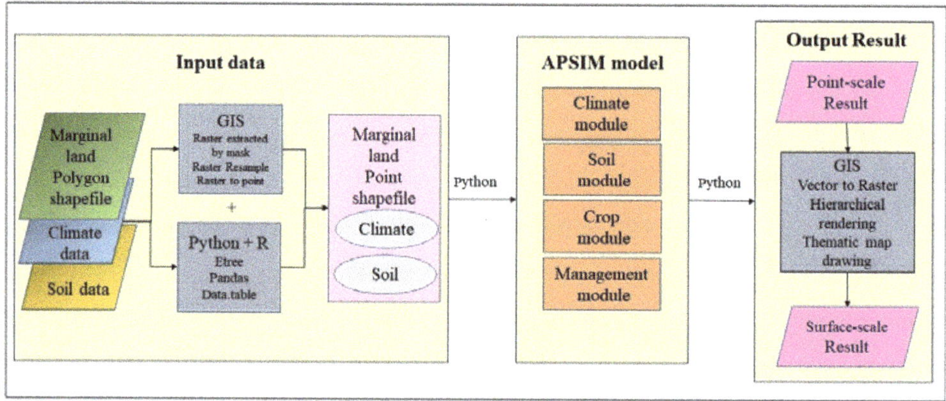


**Figure 5.** APSIM model spatial expansion technology flow chart.

*2.5. Global Sensitivity Analysis Method*

The sensitivity analysis method used in this paper was the extended Fourier amplitude sensitivity test (EFAST), which was proposed by Saltelli et al. in 2006 [30]. The algorithm is briefly introduced as follows [30]:

Model $y = f(x_1, x_2, \ldots, x_k)$ can be converted to $y = f(s)$ by an appropriate conversion function, and a Fourier transform is performed on the $y = f(s)$ as follows:

$$y = f(s) = \sum_{i=-\infty}^{\infty} \{A_j \cos(is) + B_j \sin(is)\} \tag{1}$$

$$A_i = \frac{1}{N_s} \sum_{k=1}^{N_s} f(s_k) \cos(\omega_i s_k) \tag{2}$$

$$B_i = \frac{1}{N_s} \sum_{k=1}^{N_s} f(s_k) \sin(\omega_i s_k) \tag{3}$$

where $A_i$ and $B_i$ are the Fourier amplitude, $N_s$ is the number of samples, and $i$ is the Fourier change parameter, where $i \in \bar{Z} = \left\{ -\frac{N_s-1}{2}, \ldots, -1, 0, 1, \ldots, +\frac{N_s-1}{2} \right\}$.

The spectral curve of the Fourier series is defined as $\Lambda_i = A_i^2 + B_i^2$, and by calculating the frequency $\omega_i$, the variance of the model output result caused by the change of the input parameter $x_i$ is given as follows:

$$V_i = 2 \sum_{i=1}^{+\infty} \Lambda_i \omega_i \tag{4}$$

The total variance of the model output is thus decomposed into the following:

$$V = \sum_{i=1}^{k} V_i + \sum_{i=1}^{k} \sum_{j=1}^{k} V_{ij} + \ldots + V_{i,j,\ldots,k} \tag{5}$$

where $V_i$ is the independent variance caused by the change of the $i$-th input parameter $x_i$, $V_{ij}$ is the coupling variance caused by the interaction between the $i$-th input parameter $x_i$ and the $j$-th parameter $x_j$, and by analogy, $V_{i,j,\ldots,k}$ is the variance contributed by the interaction of all input parameters, and $k$ is the number of parameters. After data normalization, the first sensitivity index $S_i$ of the parameter $x_i$ is defined as follows:

$$S_i = \frac{V_i}{V} \tag{6}$$

The total sensitivity index $S_{Ti}$ of the parameter $x_i$ is defined as follows:

$$S_{Ti} = \frac{V - V_{-i}}{V} \tag{7}$$

where $V_{-i}$ is the sum of the contribution variances of all parameters excluding the $i$-th parameter $x_i$.

## 2.6. Sensitivity Analysis

The APSIM models use many cultivar parameters to simulate crop growth. These parameters usually cannot be directly measured and need to be calibrated when the crop model is applied to a new environment or a new cultivar. Sensitivity analysis can quantify the impact of the model input parameters on the model output, thereby simplifying the calibration for new cultivars. We performed sensitivity analysis on eight crop variety parameters. The description and range of selected parameters are shown in Table 8. The variety parameters were obtained from the APSIM official website (http://www.apsim.info/) and Mao et al. [31]. The upper and lower limits of the crop variety parameters were set to be ±50% according to the default value of the model here, and all variables are subject to a uniform distribution. The additional input parameters for the model, such as the meteorological data, soil data, and crop management (e.g., seeding, fertilization) data, can be found in Section 2.2 of this paper.

**Table 8.** Upper and lower limits of sugarcane variety parameters for the sensitivity analysis.

| Name | Unit | Lower Limit | Upper Limit | Distribution |
|---|---|---|---|---|
| cane_fraction | % | 0.1 | 0.99 | Uniform |
| sucrose_delay | $g \cdot m^{-2}$ | 300 | 900 | Uniform |
| min_sstem_sucrose | $g \cdot m^{-2}$ | 750 | 2250 | Uniform |
| min_sstem_sucrose_redn | $g \cdot m^{-2}$ | 5 | 15 | Uniform |
| tt_emerg_to_begcane | $°C \cdot d$ | 1000 | 3000 | Uniform |
| tt_begcane_to_flowering | $°C \cdot d$ | 3000 | 9000 | Uniform |
| tt_flowering_to_crop_end | $°C \cdot d$ | 1000 | 3000 | Uniform |
| green_leaf_no | - | 5 | 20 | Uniform |

In this study, four outputs of the model were considered, namely, the sugarcane stem dry weight, sugarcane stem fresh weight, sucrose dry weight, and leaf area index. The specifications of the output indicators are shown in Table 9.

**Table 9.** Explanation of the selected output indicators for the parameter sensitivity analysis.

| Name | Description | Unit | Remarks |
|---|---|---|---|
| cane_wt | Sugarcane stem dry weight | $g \cdot m^{-2}$ | $1\ t \cdot ha^{-1} = 100\ g \cdot m^{-2}$ |
| canefw | Sugarcane stem fresh weight | $t \cdot ha^{-1}$ | |
| sucrose_wt | Sucrose dry weight | $g \cdot m^{-2}$ | $1\ t \cdot ha^{-1} = 100\ g \cdot m^{-2}$ |
| lai_sum | Leaf area index | $m^2 \cdot m^{-2}$ | |

We used the SIMLAB software for parameter sensitivity analysis [32]. We set the simulation number at $N = 3000$ for the sensitivity analysis in order to attain a stable convergence. Therefore, a total of 24,000 (3000 × 8) simulations were run, with eight cultivar parameters and four output indexes. The sensitivity analysis was operated using the following steps:

(1) In the statistical preprocessing module of SIMLAB software, input the range and distribution of eight crop variety parameters, use Monte Carlo method to sample all parameters 3000 times, and get the parameter sample set;

(2) Python is used to input the generated parameter sample set into the configuration file of APSIM sugarcane model, then run APSIM model from the command line, and 3000 model output results are obtained;

(3) The output of the previous step is input to the model processing module of SIMLAB, and the sensitivity analysis results of each parameter are calculated by EFAST method.

A "trial and error method" was used in this study to modify the sugarcane variety parameters. When the results of the model simulation were close to the field observation data to a greatest extent, the adjusted parameters can be set as the most appropriate parameters of the research area.

## 3. Results and Analysis

### 3.1. Sensitivity Analysis

Figure 6a shows that for cane_wt, the first three parameters with the largest first sensitivity index were green_leaf_no, cane_fraction, and tt_emerg_to_begcane, with values of 46.58%, 20.72%, and 8.16%, respectively, and the values of other parameters were less than 1%. The first three parameters with the largest global sensitivity index were the same as the first sensitivity index, with values of 68.93%, 59.22%, and 25.50%, respectively. The global sensitivity index of min_sstem_sucrose_redn was also large, ranking fourth with value of 9.52%, and the remaining parameters were less than 1%. According to Figure 6b, the sensitivity analysis results of canefw were almost the same as that of cane_wt. The first three parameters with the largest first sensitivity index were green_leaf_no, cane_fraction, tt_emerg_to_begcane, with values of 51.19%, 25.56%, and 3.85%, respectively. The first three parameters with the largest global sensitivity index were the same as the first sensitivity index, with values of 67.76%, 52.24%, and 16.97%, respectively. The global sensitivity index of min_sstem_sucrose_redn was also large, ranking fourth with value of 6.86%, and the remaining parameters were less than 1%. According to Figure 6c, for sucrose_wt, the first four parameters with the largest first sensitivity index were min_sstem_sucrose, green_leaf_no, cane_fraction, tt_emerg_to_begcane, with values of 46.27%, 11.50%, 6.51% and 2.12%, respectively, and other parameters were less than 1%. The first four parameters with the largest total sensitivity index were the same as the first sensitivity index, with values of 67.19%, 32.20%, 25.79%, 15.28%, respectively, and other parameters were less than 1%. Figure 6d shows that for lai_sum, the first three parameters with the largest first sensitivity index were cane_fraction, green_leaf_no, tt_emerg_to_begcane, with values of 60.85%, 30.18%, and 8.85%, respectively, and other parameters were less than 1%. The first three parameters with the largest global sensitivity index were the same as the first sensitivity index, with values of 65.80%, 31.62%, and 13.79%, respectively. The global sensitivity index of min_sstem_sucrose_redn ranked fourth with value of 1.71%, and the remaining parameters were less than 1%.

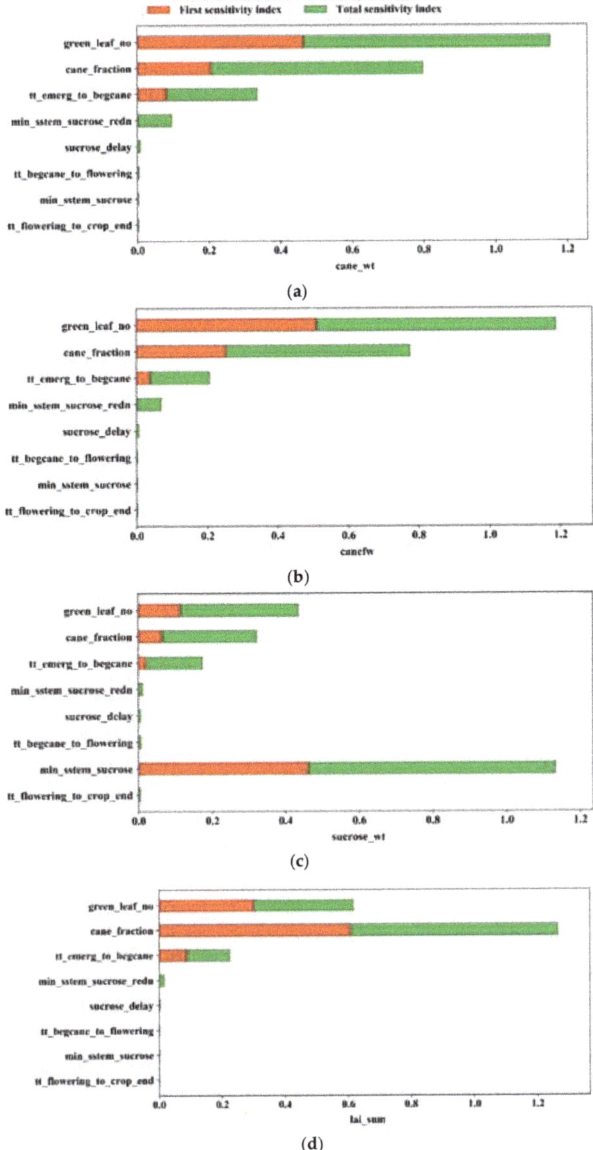

**Figure 6.** The first and total sensitivity indices for the four outputs to eight cultivar parameters in the APSIM sugarcane model. (**a–d**) Stacked bar graphs of two sensitivity indices of cane_wt, canefw, sucrose_wt, lai_sum, respectively.

In conclusion, four sugarcane yield outputs (the dry weight of the sugarcane stalk, the fresh weight of the sugarcane stalk, the dry weight of sugarcane sugar, and the leaf area index) were most sensitive to the maximum number of green leaves before plant maturation (green_leaf_no), the percentage of daily biomass allocated to cane stems (cane_fraction), and the accumulated temperature from emergence to jointing (tt_emerg_to_begcane). The dry weight of sugarcane sugar was particularly sensitive to the minimum cane stalk biomass before sugar accumulation began (min_sstem_sucrose).

### 3.2. Model Parameter Calibration

The results of the model parameter calibration based on the field observation data are shown in Table 10.

**Table 10.** Values of crop variety parameters after model parameter calibration.

| Name | New Planting [a] | Ratoon [b] |
|---|---|---|
| cane_fraction | 0.9 | 0.9 |
| sucrose_delay | 565.62 | 0 |
| min_sstem_sucrose | 1338.70 | 1500 |
| min_sstem_sucrose_redn | 15 | 10 |
| tt_emerg_to_begcane | 1700 | 1900 |
| tt_begcane_to_flowering | 5000 | 6000 |
| tt_flowering_to_crop_end | 1230.83 | 2000 |
| green_leaf_no | 90 | 90 |

[a]: The values of crop variety parameters for newly planted sugarcane after calibration. [b]: The values of crop variety parameters for ratoon sugarcane after calibration.

### 3.3. Model Validation

Figure 7 shows a linear regression graph and residual analysis graph of the observed and simulated values. The determination coefficient ($R^2$), the root mean square error (RMSE), and the consistency index (*D* index) between observed and simulated values [33] were used as statistical indicators here. In Figure 7a–c, the points in the linear regression graphs fall evenly on both sides of the 1:1 line. For the sugarcane fresh weight, emergence date and jointing date, the values of $R^2$ are 0.91, 0.76 and 0.89, the values of the RMSE are 9.32, 15.54 and 13.81, respectively, and the values of *D* are 0.97, 0.91 and 0.97. The confidence intervals of the residuals in Figure 7d–f all include zero, which indicates that the model performed correctly. In general, the model simulates the fresh weight yield of sugarcane and the phenological period well. The model validation results show that the parameters of the sugarcane varieties are ideal and that the simulation results of the APSIM model are reliable to a certain extent.

### 3.4. Model Simulation

The results for the sugarcane biomass on the marginal lands in the Guangxi Zhuang Autonomous Region of China are shown in Figure 8. The simulated sugarcane fresh weight yield in the marginal lands is ≈24.7–137.3 t·ha$^{-1}$, and the sugarcane fresh weight yield is distributed from low to high along the northeast to the southwest. However, the yield is nonuniformly distributed, which fully reflects the topography and meteorological factors. This is also in line with previous research on the climate divisions for sugarcane planting in the Guangxi Zhuang Autonomous Region [34].

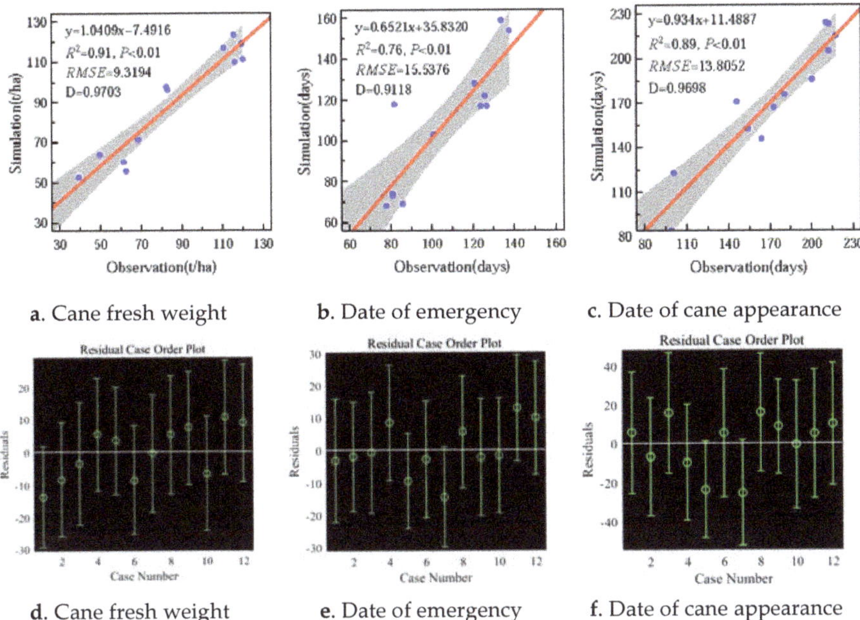

**a.** Cane fresh weight     **b.** Date of emergency     **c.** Date of cane appearance

**d.** Cane fresh weight     **e.** Date of emergency     **f.** Date of cane appearance

**Figure 7.** The linear regression diagrams ((**a**) cane fresh weight; (**b**) date of emergency; (**c**) date of cane appearance) and the residual analysis diagrams ((**d**) cane fresh weight; (**e**) date of emergency; (**f**) date of cane appearance) of the observed and simulated values.

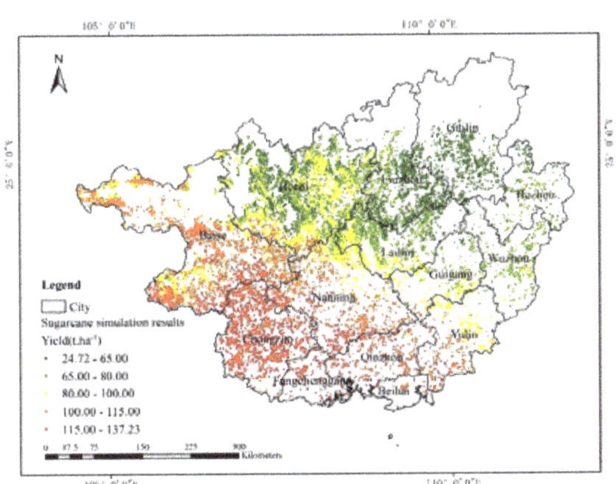

**Figure 8.** Spatial distribution of simulated sugarcane yield on the marginal lands in the Guangxi Zhuang Autonomous Region.

*3.5. Fuel Ethanol Production Potential*

The relevant literature shows that the conversion rate of sugarcane stem yield to ethanol fuel production is 12:1 [35,36] and that 50% of sugarcane produced in cropland can be used for ethanol production after meeting the demands of the sugar industry [37]. Therefore, the spatial distribution of sugarcane ethanol production could be estimated according to the administrative divisions of the Guangxi Zhuang Autonomous Region, as shown in Table 11. The total sugarcane output of Guangxi Zhuang Autonomous Region can be obtained by summing the sugarcane produced on the marginal land and cropland, as shown in Figure 9. Table 11 shows that the total area of marginal land in the Guangxi Zhuang Autonomous Region is 53,124 km$^2$, accounting for 22.36% of the total area. The total yield of sugarcane on marginal land is 42,522.05 $\times$ 10$^4$ t. The total yield of ethanol is estimated to be 3847.37 $\times$ 10$^4$ t when combining marginal land ethanol with cropland ethanol. According to the Guangxi Statistical Yearbook [19], the number of cars owned in Guangxi is 590.4 $\times$ 10$^4$, the consumption of gasoline is 392.29 $\times$ 10$^4$ t, and replacing 10% of gasoline with ethanol fuel requires 392.29 $\times$ 10$^3$ t of ethanol fuel. After deducting vehicle ethanol fuel, there are still 3808.14 $\times$ 10$^4$ t of ethanol fuel that can be exported to the ASEAN.

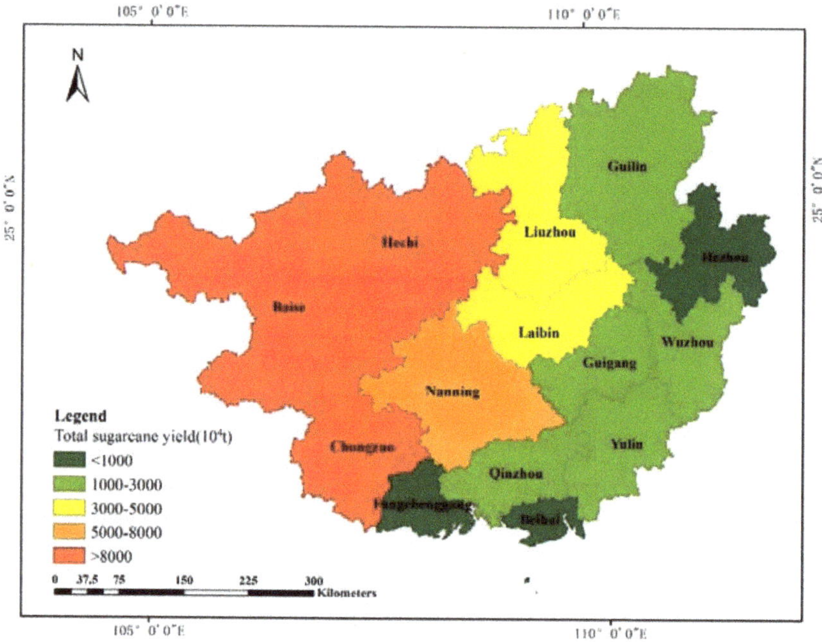


**Figure 9.** Total sugarcane yield in the Guangxi Zhuang Autonomous Region.

**Table 11.** Statistics of sugarcane-based ethanol energy potential in cities of Guangxi Zhuang Autonomous Region.

| District | District Area $(km^2)$ | Marginal Land Area $(km^2)$ | Sugarcane Stem Fresh Weight Yield $(10^4\ t)$ | Sugarcane Yield of Cropland $(10^4\ t)$ | Ethanol from Marginal Land $(10^4\ t)$ | Ethanol from Cropland $(10^4\ t)$ | Total Ethanol $(10^4\ t)$ |
|---|---|---|---|---|---|---|---|
| Baise | 36,201 | 9188 | 9771.00 | 257.29 | 814.25 | 10.72 | 824.97 |
| Beihai | 3989 | 1984 | 142.08 | 236.14 | 11.84 | 9.84 | 21.68 |
| Chongzuo | 17,332 | 4616 | 5613.28 | 2558.86 | 467.77 | 106.62 | 574.39 |
| Fangchenggang | 6239 | 556 | 688.30 | 301.08 | 57.36 | 12.55 | 69.90 |
| Guigang | 10,602 | 1248 | 997.06 | 227.40 | 83.09 | 9.48 | 92.56 |
| Guilin | 27,667 | 2628 | 1412.89 | 29.06 | 117.74 | 1.21 | 118.95 |
| Hechi | 33,476 | 12,232 | 9104.59 | 296.07 | 758.72 | 12.34 | 771.05 |
| Hezhou | 11,753 | 4108 | 739.33 | 12.41 | 61.61 | 0.52 | 62.13 |
| Laibin | 13,382 | 3932 | 2853.74 | 1108.72 | 237.81 | 46.20 | 284.01 |
| Liuzhou | 18,597 | 4108 | 2537.78 | 618.33 | 211.48 | 25.76 | 237.25 |
| Nangling | 22,099 | 4368 | 4646.96 | 1163.60 | 387.25 | 48.48 | 435.73 |
| Qinzhou | 10,897 | 1280 | 1492.07 | 317.08 | 124.34 | 13.21 | 137.55 |
| Wuzhou | 12,572 | 1372 | 996.57 | 11.02 | 83.05 | 0.46 | 83.51 |
| Yulin | 12,824 | 1504 | 1526.38 | 155.69 | 127.20 | 6.49 | 133.69 |
| Total | 237,630 | 53,124 | 42,522.05 | 7292.76 | 3543.50 | 303.87 | 3847.37 |

Note: The first column is the subordinate city name in Guangxi. The second column is the area of each city. The third column is the marginal land area in each city. The fourth column is the sugarcane cane yield from marginal land in each city as calculated by the ArcGIS software. The fifth column is the sugarcane yield from cropland in Guangxi, as per the Guangxi Statistical Yearbook. The sixth column is the estimated ethanol of sugarcane planted on the marginal land, which is obtained by multiplying the sugarcane yield by the sugarcane ethanol conversion rate. The seventh column is the estimated ethanol of sugarcane planted on the farmland, which is obtained by multiplying 50% of the sugarcane yield by the sugarcane ethanol conversion rate. The eighth column is the estimated total ethanol of each city, which is obtained by summing the sixth and seventh columns.

## 4. Discussion

Guangxi is a pilot region in China that vigorously promotes the development of the non-grain biomass energy industry [38]. The results of this study show that the average yield of sugarcane on marginal land is about 80 t·ha$^{-1}$, and that on cultivated land is about 82 t·ha$^{-1}$ (calculated by dividing the total output by the total area), which indicates that Guangxi has great potential for planting sugarcane as an energy crop, as Guangxi has a large amount of unused shrub forest lands, sparse forest lands, and grasslands suitable for planting sugarcane [39], and the local climate is also suitable. Moreover, the distribution of sugarcane production in marginal land in Guangxi (Figure 8) indicates that the sugarcane production potential in marginal land in southwestern Guangxi is large. Considering the scattered distribution of unused land, ecological safety, and other factors, the development of sugarcane as a non-grain biomass energy should be prioritized in the marginal land in the west and southwest.

This paper uses a surface-to-point method to predict sugarcane yield in marginal land by combining the APSIM sugarcane model and GIS spatial analysis technology, which not only overcomes the difficulty of unpredictable yields in marginal land but also considers the hydrothermal conditions and physiological characteristics of sugarcane when compared with other research [24,40–42]. However, there are some limitations that need to be addressed in this study. First, due to the vast amount of land and the varying environmental conditions in the study area, the crop growth model was verified by only a limited number of sites, and the accuracy of the parameters thus cannot be ensured. Secondly, this study has not taken into account all the impact factors on the growth and development of sugarcane in the APSIM sugarcane model, such as the impact of pests and other disasters caused by meteorological factors in the real production process. In addition, this paper creates a spatial resolution problem when the GIS spatially expands the crop growth model. It is unknown at what resolution the model can be extended spatially and at what resolution the effect is best. Therefore, the feasibility and optimal parameters for the spatial expansion of the crop growth model should be further studied. Furthermore, regarding the feasibility of planting sugarcane on marginal land, due to the high planting density, long growing period, poor soil conditions of the planting area, and the large amount of rainfall in Guangxi, the large-scale utilization of idle forest land, grassland, and other unused land to plant sugarcane can cause serious soil and water loss, and even lead to the risk of landslides over time. Moreover, if sugarcane is planted on a large scale, the lack of adequate management measures will inevitably lead to a decline in soil fertility and thus a decline in soil production potential. Therefore, it is necessary to further evaluate the impact of sugarcane cultivation on the ecological environment in subsequent studies.

## 5. Conclusions

This study assessed the production potential of sugarcane as an energy crop in the Guangxi Zhuang Autonomous Region. First, the marginal land resources suitable for sugarcane were extracted. Next, a sensitivity analysis, a calibration, and a verification of the APSIM model were carried out to confirm its applicability. Finally, the growth process of sugarcane was simulated for the study area.

The results show that the APSIM sugarcane model simulates the sugarcane stem yield and phenological period of sugarcane in Guangxi Zhuang Autonomous Region well, and that the related statistical graphics and indicators also perform well. Assuming that sugarcane is planted as an energy crop on the marginal lands of the study region, approximately 42,522.05 × 10$^4$ t of sugarcane stems can be harvested each year. It is estimated that the sugarcane produced on the marginal land plus 50% of the sugarcane from croplands can produce about 3847.37 × 10$^4$ t of ethanol fuel. After meeting the demands for vehicle ethanol fuel in Guangxi, 3808.14 × 10$^4$ t of ethanol fuel remain that can be exported to the ASEAN.

Due to the many uncertainties mentioned in Section 4, the next step should be to consider more factors that affect the sustainable development of sugarcane bioenergy, including ecological security, the technology to produce bioenergy from sugarcane, energy efficiency, and the increase of greenhouse gas efficiency. In general, the sustainable development of sugarcane bioenergy should be analyzed in conjunction with life cycle assessments and biogeochemical process models.

**Author Contributions:** Conceptualization, J.F. and D.J.; Data curation, J.D.; Investigation, J.D.; Methodology, T.P. and J.F.; Project administration, D.J.; Resources, J.D.; Writing–original draft, T.P.; Writing–review & editing, J.F. and D.J. All authors have read and agreed to the published version of the manuscript.

**Funding:** This work was supported by National Natural Science Foundation of China (Grant No. 41971250), Youth Innovation Promotion Association (Grant No. 2018068).

**Acknowledgments:** We greatly thank "MDPI English editing" (English-42266) for the editing assistance.

**Conflicts of Interest:** The authors declare no conflict of interest.

## References


1. Tian, C. Analysis of China's Oil and Gas Import and Export in 2013. *China Oil Gas* **2014**, *21*, 36–41.
2. Liu, Q.; Gu, A.; Teng, F.; Song, R.; Chen, Y. Peaking China's $CO_2$ Emissions: Trends to 2030 and Mitigation Potential. *Energies* **2017**, *10*, 209. [CrossRef]
3. Chen, W.-H.; Lee, K.T.; Ong, H.C. Biofuel and Bioenergy Technology. *Energies* **2019**, *12*, 290. [CrossRef]
4. Han, Y.; Kagawa, S.; Nagashima, F.; Nansai, K. Sources of China's Fossil Energy-Use Change. *Energies* **2019**, *12*, 699. [CrossRef]
5. National Energy Administration. *Implementation Plan on Expanding Biofuel Ethanol Production and Promoting the Use of Automobile Ethanol Gasoline*; National Energy Administration: Beijing, China, 2017.
6. National Development and Reform Commission. *"Eleventh Five-Year Plan" Development Special Plan for Biofuel Ethanol and Automobile Ethanol Gasoline*; National Development and Reform Commission: Beijing, China, 2007.
7. Ministry of Commerce. *China-ASEAN Comprehensive Economic Cooperation Framework Agreement*; Ministry of Commerce: Beijing, China, 2002.
8. Li, Y.; Hou, H.; Wang, X.; Kuang, Z. Research Status and Prospect of Energy Plant. *J. Anhui Agric.* **2013**, *41*, 1682–1683.
9. Shi, S.; Cheng, D.; Ma, F. Exploitation and Utilization of Biomass Energy Crop-Energy Beet. *Chin. Agric. Sci. Bull.* **2007**, *23*, 416–419.
10. Xie, L.; Li, P.; Zhang, W.; Yang, M. A review on the development potential of bioenergy rapeseed. *Chin. J. Bioprocess Eng.* **2005**, *1*, 32–35.
11. Cervi, W.R.; Camargo Lamparelli, R.A.; Abel Seabra, J.E.; Junginger, M.; van der Hilst, F. Bioelectricity potential from ecologically available sugarcane straw in Brazil: A spatially explicit assessment. *Biomass Bioenergy* **2019**, *122*, 391–399. [CrossRef]
12. Singels, A.; Jarmain, C.; Bastidas-Obando, E.; Olivier, F.C.; Paraskevopoulos, A.L. Monitoring water use efficiency of irrigated sugarcane production in Mpumalanga, South Africa, using SEBAL. *Water SA* **2018**, *44*, 636–646. [CrossRef]
13. Yawson, D.O.; Adu, M.O.; Osei, K.N. Spatial assessment of sugarcane (*Saccharurn spp.* L.) production to feed the Komenda Sugar Factory, Ghana. *Heliyon* **2018**, *4*, e00903. [CrossRef]
14. Lisboa, I.P.; Damian, J.M.; Cherubin, M.R.; Silva Barros, P.P.; Fiorio, P.R.; Cerri, C.C.; Pellegrino Cerri, C.E. Prediction of Sugarcane Yield Based on NDVI and Concentration of Leaf-Tissue Nutrients in Fields Managed with Straw Removal. *Agronomy* **2018**, *8*, 196. [CrossRef]
15. Satiro, L.S.; Cherubin, M.R.; Lisboa, I.P.; Noia Junior, R.S.; Cerri, C.C.; Pellegrino Cerri, C.E. Prediction of Sugarcane Yield by Soil Attributes under Straw Removal Management. *Agron. J.* **2019**, *111*, 14–23. [CrossRef]
16. Dias, H.B.; Sentelhas, P.C. Sugarcane yield gap analysis in Brazil-A multi-model approach for determining magnitudes and causes. *Sci. Total Environ.* **2018**, *637*, 1127–1136. [CrossRef] [PubMed]
17. Rodriguez, R.G.; Scanlon, B.R.; King, C.W.; Scarpare, F.V.; Xavier, A.C.; Pruski, F.F. Biofuel-water-land nexus in the last agricultural frontier region of the Brazilian Cerrado. *Appl. Energy* **2018**, *231*, 1330–1345. [CrossRef]
18. Sanches, G.M.; Nonato de Paula, M.T.; Graziano Magalhaes, P.S.; Duft, D.G.; Vitti, A.C.; Kolln, O.T.; Montes Nogueira Borges, B.M.; Junqueira Franco, H.C. Precision production environments for sugarcane fields. *Sci. Agric.* **2019**, *76*, 10–17. [CrossRef]

19. Guangxi Zhuang Autonomous Region Statistics Bureau. *Guangxi Statistical Yearbook*; Guangxi Zhuang Autonomous Region Statistics Bureau: Beijing, China, 2019.

20. Huang, X.; Zhou, H.; Huang, M.; Zhao, J. Climate and temperature changes in Guangxi in the past 50 years. *Guangxi Meteorol.* **2005**, *26*, 9–11.

21. Ministry of Agriculture and Rural Affairs. *Opinions of the Ministry of Agriculture on Strengthening Energy Conservation and Emission Reduction in Agriculture and Rural Areas*; Ministry of Agriculture and Rural Affairs: Beijing, China, 2007.

22. Fu, J.; Jiang, D.; Hao, M. *Research on the Development Potential of Non-Grain Fuel Ethanol in China*; China Meteorological Press: Beijing, China, 2017.

23. Li, R.; Zhang, Y.; Yang, D.; Qu, Y.; Guo, J. Study on development potential of full mechanization in diversity terrain of Yunnan sugarcane region. *Chin. Agric. Mech.* **2012**, *4*, 71–74.

24. Ruan, H. Simulation Study on the Impact of Climate Change on Sugarcane Production Potential in Guangxi. Ph.D. Thesis, Guangxi University, Guangxi, China, 2018.

25. He, H.; Luo, C.; Tang, L.; Ma, W.; Chen, H.; Wei, H.; Huang, Z. High-yield Cultivation Techniques of Guitang 32 in Chongzuo Cane Area. *Chin. Trop. Agric.* **2016**, *3*, 69–71.

26. Zu, Q.; Mi, C.; Liu, D.L.; He, L.; Kuang, Z.; Fang, Q.; Ramp, D.; Li, L.; Wang, B.; Chen, Y.; et al. Spatio-temporal distribution of sugarcane potential yields and yield gaps in Southern China. *Eur. J. Agron.* **2018**, *92*, 72–83. [CrossRef]

27. Xu, T.; Hutchinson, M.F. New developments and applications in the ANUCLIM spatial climatic and bioclimatic modelling package. *Environ. Model. Softw.* **2013**, *40*, 267–279. [CrossRef]

28. Dai, Y.; Wei, S.; Duan, Q.; Liu, B.; Niu, G. Development of a China Dataset of Soil Hydraulic Parameters Using Pedotransfer Functions for Land Surface Modeling. *J. Hydrometeorol.* **2013**, *14*, 869–887. [CrossRef]

29. Zuur, A.F.; Ieno, E.N.; Meesters, E.H.W.G. *A Beginner's Guide to R*; Springer: New York, NY, USA, 2009.

30. Saltelli, A.; Ratto, M.; Tarantola, S.; Campolongo, F. Sensitivity analysis practices: Strategies for model-based inference. *Reliab. Eng. Syst. Saf.* **2006**, *91*, 1109–1125. [CrossRef]

31. Mao, J.; Inman-Bamber, N.G.; Yang, K.; Lu, X.; Liu, J.; Jackson, P.A.; Fan, Y. Modular Design and Application of Agricultural System Simulating Model for Sugarcane (APSIM-Sugar). *Sugar Crop. China* **2017**, *39*, 44–50.

32. Tarantola, S.; Becker, W. Simlab 4.0 for Global Sensitivity Analysis. In *Handbook of Uncertainty Quantification*; Springer: New York, NY, USA, 2016; pp. 1–21.

33. Willmott, J.C. Some Comments on the Evaluation of Model Performance. *Bull. Am. Meteorol. Soc.* **1982**, *63*, 1309–1313. [CrossRef]

34. Su, Y.; Li, Z.; Sun, H. Climate Division of Sugarcane Planting Based on GIS in Guangxi. *Chin. J. Agrometeorol.* **2006**, *27*, 252–255.

35. Li, Q.; Qi, R.; Zhang, Y. Development prospects of energy sugarcane to produce fuel ethanol. *Sugarcane Canesugar* **2004**, *5*, 29–33.

36. Zeng, L.; An, Y.; Li, Q. Technical, Economic and Environmental Analysis of Fuel Ethanol Production from Sugarcane in China. *Sugarcane Canesugar* **2006**, *2*, 15–19.

37. Li, Y.; Tan, Y.; Li, S.; Yang, R. Analyses on the potential of sugarcane as a bioenergy crop in China. *Southwest China J. Agric. Sci.* **2006**, *19*, 742–746.

38. Xie, G. Progress and direction of non-food biomass feedstock supply research and development in China. *J. China Agric. Univ.* **2012**, *17*, 1–19.

39. Zhang, J. Land potential analysis of biomass energy development in Guangxi. *Mark. Forum* **2008**, *2*, 9–14.

40. Xu, Y.; Zhang, W. The Application on the Sugar Cane Forecast with the Neural Networks Using Improved Genetic Algorithms. *J. South China Agric. Univ.* **2010**, *31*, 102–104.

41. Mao, J.; Wang, J.; Huang, M.; Lu, X.; Dao, J.; Zhang, Y.; Tao, L.; Yu, H. Effects of sowing date, water and nitrogen coupling management on cane yield and sugar content in sugarcane region of Yunan. *Trans. Chin. Soc. Agric. Eng.* **2019**, *35*, 134–144.

42. Peng, Q.; Feng, L.; Deng, J.; Fan, X.; Zhang, Y. Application of BP Neural Network in Predication of Sugarcane Yield in Yunnan Province. *Sugar Crop. China* **2019**, *41*, 54–57.

Article

# Coconut Wastes as Bioresource for Sustainable Energy: Quantifying Wastes, Calorific Values and Emissions in Ghana

George Yaw Obeng [1,*], Derrick Yeboah Amoah [2], Richard Opoku [1], Charles K.K. Sekyere [1], Eunice Akyereko Adjei [1] and Ebenezer Mensah [2]

[1] Department of Mechanical Engineering, College of Engineering, Kwame Nkrumah University of Science and Technology, Kumasi, Ghana; ropoku.coe@knust.edu.gh (R.O.); ckksekyere@knust.edu.gh (C.K.K.S.); eadjeiakyereko.coe@knust.edu.gh (E.A.A.)

[2] Department of Agricultural and Biosystems Engineering, College of Engineering, Kwame Nkrumah University of Science and Technology, Kumasi, Ghana; derrickyeboah@hotmail.com (D.Y.A.); emensah.coe@knust.edu.gh (E.M.)

* Correspondence: gyobeng.coe@knust.edu.gh

Received: 7 March 2020; Accepted: 26 March 2020; Published: 1 May 2020

**Abstract:** Coconut husks with the shells attached are potential bioenergy resources for fuel-constrained communities in Ghana. In spite of their energy potential, coconut husks and shells are thrown away or burned raw resulting in poor sanitation and environmental pollution. This study focuses on quantifying the waste proportions, calorific values and pollutant emissions from the burning of raw uncharred and charred coconut wastes in Ghana. Fifty fresh coconuts were randomly sampled, fresh coconut waste samples were sun-dried up to 18 days, and a top-lit updraft biochar unit was used to produce biochar for the study. The heat contents of the coconut waste samples and emissions were determined. From the results, 62–65% of the whole coconut fruit can be generated as wastes. The calorific value of charred coconut wastes was 42% higher than the uncharred coconut wastes. $PM_{2.5}$ and CO emissions were higher than the WHO 24 h air quality guidelines (AQG) value at 25 °C, 1 atmosphere, but the CO concentrations met the WHO standards based on exposure time of 15 min to 8 h. Thus, to effectively utilise coconut wastes as sustainable bioresource-based fuel in Ghana, there is the need to switch from open burning to biocharing in a controlled system to maximise the calorific value and minimise smoke emissions.

**Keywords:** coconut wastes; bioenergy resource; pollutant emissions; calorific value; biocharing

---

## 1. Introduction

Coconut is a perennial fruit that thrives well on sandy soils and mostly grows well on islands and coastal areas in the tropics and rainforest climate, especially along the coastline zones where it enjoys the sun irradiation as well as water [1]. Globally, several million tonnes of coconut are produced annually in Asia, Latin America and Africa. As of the year 2018, the total world production of coconut was 250–300 million tonnes [2]. Every part of the coconut plant is useful with a wide range of products being obtained from it [3–6]. Fresh coconut fruit is appreciated for its juice, food and animal feed; coconut husks are used as raw material supply [7–12] and wall hangings; fibres are used for clothing and bags, among other uses [13]. The shell normally takes a long time to decompose and often becomes a nuisance. Coconut husks with the shells attached and other biomaterials including straw, rice husks, corn stalks, sawdust, cereal husks, sugarcane bagasse and nutshells are a potential bioresource that can be used as domestic fuel [14] in energy-poor communities, such as those found in Ghana where about 73% of households depend on firewood for cooking and water heating [15].

In Ghana, after the edible portions of coconut fruits are consumed, the wastes in the form of husks and shells are usually thrown away or openly burned. The problem is that open burning and improper throwing away of coconut wastes (husks and shells) result in poor sanitation, air pollution and blocked roadside drains that facilitate the breeding of mosquitoes. Local food vendors use either the raw unprocessed coconut wastes or dry them in the open sun for a number of days to reduce the moisture content before employing as fuel for domestic cooking. Lee and Park [16] reported that inefficient combustion of biomass can release a considerable amount of various airborne pollutants, including particulates and carbon monoxide. Exposure to varying concentrations of pollutant emissions can affect people's health as well as the environment. It is reported that exposure to ultra-fine particulates ($PM_{01}$–$PM_{2.5}$) could increase the risk of severe respiratory diseases [17].

In some Ghanaian communities, coconut sellers at times persuade food vendors to collect the wastes free of charge for use as an alternative to firewood. In the light of these problems, there is the need for continuous research in order to gain insight into the quantity of whole coconut that can be generated as waste, the caloric value and the resulting pollutant emissions.

Parametric data and findings of this study will significantly contribute to the knowledge of the necessity to locally innovate in systems and processes that can be effectively utilised to optimise the waste-to-energy process so as to reach the goal of clean bioenergy production with low carbon emissions. The results of this study will provide data that can be used to estimate the amount of energy that can be produced from known quantities of coconut wastes and several other bioresources such as straw, rice husks, corn stalks, sawdust, cereal husks, sugarcane bagasse, nutshells etc. Such bioresources can be efficiently converted to produce clean biochar briquette fuels for heat and electricity generation in fuel-constraint communities [14,15].

Consequently, this study seeks to achieve the following objectives: 1) quantify the amount of wastes that can be generated from whole coconut; 2) determine the calorific values of raw uncharred and charred coconut wastes; and 3) analyse the moisture content and the resulting carbon monoxide (CO) and particulate matter ($PM_{2.5}$) emissions from the burning of coconut waste for possible emission reduction measures to improve the quality of combustion.

## 2. Materials and Methods

### 2.1. Quantifying the Proportion of Waste to be Generated from Whole Coconut

To quantify the proportion of waste that can be generated from whole coconut, fresh coconuts were purchased from a local dealer who obtains his coconuts from the westernmost district located on the coast of Ghana, known as the Jomoro district. In this study, both the pure coconut breed referred to as the local variety and mixed breed referred to as hybrid variety were used. Using an electronic weighing scale CTS 3000 (with 1 g minimum accuracy), a random sample of 50 whole coconuts were weighed to collect data on the individual weights. The fresh coconuts were dehusked using a machete as shown in Figure 1. They were individually weighed to obtain the quantity of husks by weight. The individual shells were also removed, weighed and recorded as quantity of shells by weight. Figure 2 is a half view of a whole coconut showing the skin, husk, shell and copra.

The weight of the fruit (juice and copra) was obtained by finding the difference between the weight of the whole coconut fruit and weight of husk and shell. The percentage composition of the coconut waste (husk and shell) by weight was determined using the formula in Equation (1):

$$\text{Percentage waste} = \frac{\text{sum of waste (husk + shell) weight}}{\text{total weight}} \times 100\% \tag{1}$$

**Figure 1.** Picture showing the dehusking of fresh coconut fruit.

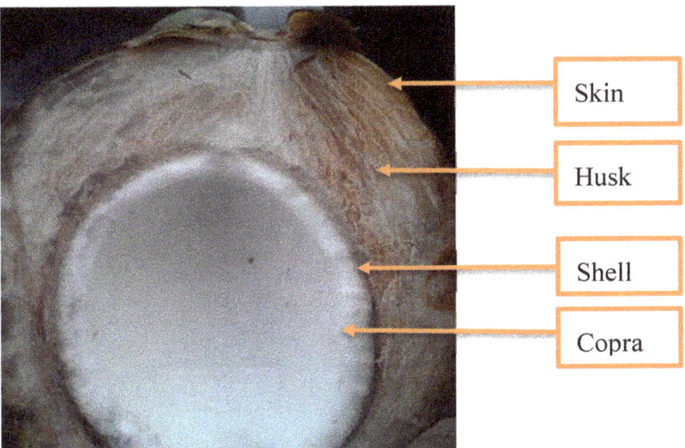


**Figure 2.** Half view of whole coconut showing the husk, skin, shell and copra.

*2.2. Drying and Determination of Moisture Content*

The oven-drying method was used to compare the moisture content of the sun-dried coconut wastes that were determined using the pin-type moisture meter (J-2000 Delmhost Instrument type with accuracy of ± 0.2). All things being equal, the moisture contents were determined to understand the influence of moisture content on pollutant emissions produced when coconut wastes are burned raw at the local community level.

Samples of the coconut wastes (husks and shells) were sun-dried for 3 to 18 days. Using the pin-type moisture meter, the moisture contents of randomly sampled coconut wastes were measured for 3 to 18 days as shown in Figure 3.

**Figure 3.** Field measurement of moisture content using the moisture meter.

Samples of about 200 g of the coconut wastes were also measured using the electronic weighing scale CTS 3000. The coconut wastes samples were then dried in an oven at a temperature of 105 °C for 24 h, then the samples of coconut wastes were weighed, and the moisture losses were determined by subtracting the oven-dry weight from the moist weight. The moisture content (Mc) of the coconut waste samples was determined as the mass of water in the sample expressed as a percentage of the dry mass as shown in Equation (2).

$$\text{Moisture content, Mc} = \frac{MW - MD}{MW} \times 100 \ (\%) \tag{2}$$

where, MW = wet weight and MD = dry weight

*2.3. Charring*

The charring experiment was carried out at the Food Processing Unit of the Technology Consultancy Centre, Kwame Nkrumah University of Science and Technology, Kumasi, Ghana. A top-lit updraft (TLUD) biochar unit of a metallic drum of dimensions (Ø57cm x 85 cm high) was used. It had a chimney dimension of Ø21 cm x 120 cm high attached to a metal lid of Ø54.5cm x 25 cm high as shown in Figure 4. Holes were perforated beneath the reactor and it was mounted on three stones to enhance air flow, while ensuring it is stable. Then 5 kg of coconut waste sample was weighed and poured into the reactor ensuring that the coconut wastes are spread out evenly. A handful of dried leaves were used to kindle the fire from the top to start the combustion process. The metal lid with chimney was then fitted onto the reactor container to stop further entrance of oxygen as well as to provide a channel for the smoke to escape. Temperature of the container was recorded at regular intervals of time using an infrared thermometer. The temperature values measured ranged from 74.2 °C to 406.8 °C. In order to ascertain that the process was complete, drops of water were thrown on the side of the reactor container, when instantaneous puffs of steam close to the bottom were observed, the process was then considered to be completed. The charring test was repeated three times and the average values of the variables were determined. The coconut waste samples were reduced into smaller pieces and crashed using a hammer mill. The milled samples were then sieved using the Tyler sieves to obtain the appropriate particle sizes.

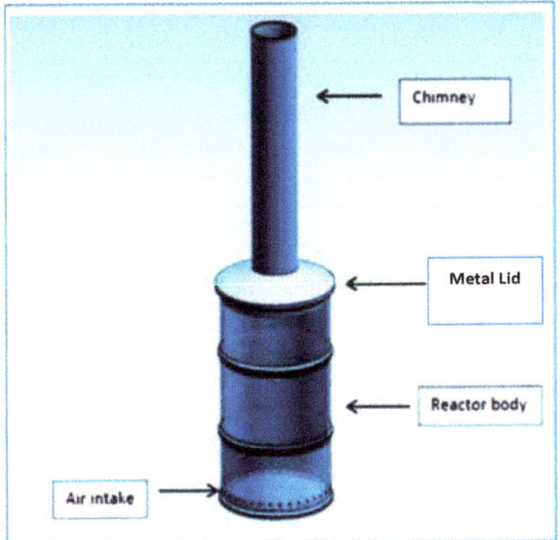


**Figure 4.** Schematic of top-lit updraft (TLUD) biochar unit.

## 2.4. Determination of Calorific Value

In this study, a bomb calorimeter SDC311 was used to determine the heat content of the coconut wastes. The bomb calorimeter conforms to ASTMD5865 standard. The specifications of the bomb calorimeter include analysis time of 11 min; precision of RSD< 0.1%; oxygen gas requirement of 99.5% purity etc. The crucible in the bomb calorimeter was placed on the weighing pan of the analytical balance to measure its weight. Using the prongs, one gram of the sample was fetched into the crucible on the analytical balance; the crucible was placed onto the crucible support of the oxygen bomb. Both ends of the firing wire were connected to two electrode rods of the oxygen bomb by bending them in a circular manner for firm contact.

Thereafter, the oxygen bomb core was moved into the oxygen bomb cylinder that had been filled with 10 mL of distilled water earlier on. After that, the oxygen bomb cover was tightly closed. Next, the oxygen bomb was filled with oxygen to about 2.8 to 3.0 MPa of pressure. The oxygen bomb was immersed into a bucket of water to determine the presence of leakage. Being satisfied with the outcome, the oxygen bomb was placed inside the bomb calorimeter and closed, then the system automatically begun the test. After about 10 min when the test was completed, the sample was completely combusted. The bomb calorimeter is instrumented such that after complete combustion of the sample, the calorific value is computed and displayed by running software on the windows-based desktop computer. After taking the readings, the calorimeter was opened to take the sample out. In doing this, oxygen was released using a release valve and then, the crucible taken out, washed in distilled water and cleaned with the bomb towel. To determine the calorific value, the experiment was conducted three times and the average calorific values were computed. The experiment was conducted at the Cookstove Testing and Expertise Laboratory (C-Lab) of the Kwame Nkrumah University of Science and Technology (KNUST), Kumasi, Ghana.

## 2.5. Determination of Emissions

An indoor air pollution meter (IAP meter 500 series) was used for measuring the emissions. The resulting carbon monoxide (CO) and particulate matter ($PM_{2.5}$) measures (with a site-specific gravimetric calibration) provided an assessment of exposure to emissions. Relative humidity of 60–73% and ambient temperature of 30–34 °C were recorded during the test.

The indoor air pollution meter (IAP 5019)-Aprovecho Research Centre model was used for the emission measurements.

Before the experiment began, the IAP was opened for about 15 min to allow it to get accustomed to the local temperature since the CO sensor is very sensitive to temperature. Slow mode sampling rate was selected owing to the duration of the experiment. Thereafter, the meter was switched on for one hour to activate the system. The IAP was then hung up at the charring site, and the smoke produced from the coconut wastes that were sun-dried for 3 and 15 days is shown in Figure 5a,b. The different time periods were measured. When the charring process began, a few minutes were allowed to elapse to allow the burning to start up well devoid of unnecessary smoke, before timing as "test begins". After charring, the meter was switched off and the time was recorded as the test ends. The IAP was equipped with an SD card that stored the measured data. The data were then processed on a computer using software programmes such as Terreterm and Livegraph for connecting the meter directly to the computer.

(a)            (b)

**Figure 5.** Smoke emissions from coconut wastes sun-dried for (**a**) 3 days and (**b**) 15 days.

## 3. Results and Discussion

### 3.1. Proportion of Whole Coconut Waste

The range, mean and standard deviations of the weights of the whole coconut fruit, husks, shells and copra/juice of the 50 samples of both hybrid and local varieties are presented in Table 1. From the results, an average husk weight of 0.80 ± 0.14 kg and shell weight of 0.25 ± 0.08 kg were determined for the hybrid variety with a total weight of 1.68 ± 0.21 kg. The proportion by weight of the waste husks and shells of the hybrid variety amounted to 62.62% of the whole coconut fruit. In Table 1, an average husk weight of 1.12 ± 0.33 kg and shell weight of 0.34 ± 0.09 kg were determined for the local variety with a total weight of 2.23 ± 0.61 kg. The proportion by weight of both the waste husks and shells of the local variety was about 65.60% of the whole coconut. Overall, a whole coconut fruit can yield husk waste of 47–50% and shell waste of 14–15%. The study results also revealed that 62–66% of the whole coconut is likely to be generated as husk and shell wastes, which can be considered as useful bioresource for sustainable energy production in fuel-constrained communities.

**Table 1.** Measured values on weight of whole coconut, husks, shells and copra/juice of hybrid and local coconut varieties.

| | Whole Coconut Fruit | Husk | Shell | Copra and Juice |
|---|---|---|---|---|
| **Hybrid Coconut Variety** | | | | |
| Range (min–max) (kg) | 1.29–2.11 | 0.57–1.09 | 0.13–0.44 | 0.32–0.97 |
| Mean weight + Std. dev (kg) | 1.68 ± 0.21 | 0.80 ± 0.14 | 0.25 ± 0.08 | 0.53 ± 0.18 |
| Weight proportion (%) | 100 | 47.75 | 14.87 | 37.38 |
| Husk + shell weight (%) | | 62.62 | | |
| Coefficient of variation (%) | 12.5 | 17.5 | 32 | 33.96 |
| **Local Coconut Variety** | | | | |
| Range (min–max) (kg) | 1.51–3.53 | 0.69–2.08 | 0.20–0.56 | 0.34–1.55 |
| Mean weight + Std. dev (kg) | 2.23 ± 0.61 | 1.12 ± 0.33 | 0.34 ± 0.09 | 0.77 ± 0.29 |
| Weight proportion (%) | 100 | 50.16 | 15.44 | 34.41 |
| Husk + shell weight (%) | | 65.60 | | |
| Coefficient of variation (%) | 26.35 | 29.46 | 26.47 | 37.66 |
| Sample size ($N_1$ = hybrid variety; $N_2$ = local variety) | | $N_1$ = 25; $N_2$ = 25 | | |

## 3.2. Variability in the Various Parts

To find the variation in sizes in regard to the mean weights, coefficient of variation (CV) was used. CV is the ratio of sample standard deviation to the sample mean. According to Kelly and Donnelly [18], lower CV values are more consistent than higher CV values. In Table 1, the hybrid coconut variety shows lower CV values than the local coconut variety, indicating there is less variation in the size of the hybrid coconut variety than the local variety.

Further, Figure 6 depicts graphs that show variability in the weights of both local and hybrid coconut varieties. The trendlines provide a vivid picture and graphical representation of the variability in the weight of the coconuts. It is observed that there is relatively less variation in the weights of the hybrid coconut variety than the local variety. A relatively low degree of variation would mean better uniformity or consistency in the sizes of the hybrid coconut variety. What it means is that the dataset on the local variety of coconut contains values considerably higher and lower than their mean weight when compared to the dataset on the hybrid variety of coconut. In general, coconut hybrids are much preferred by coconut growers. Hence, different forms of varieties are exploited as breeding materials for coconut hybrid production [19].

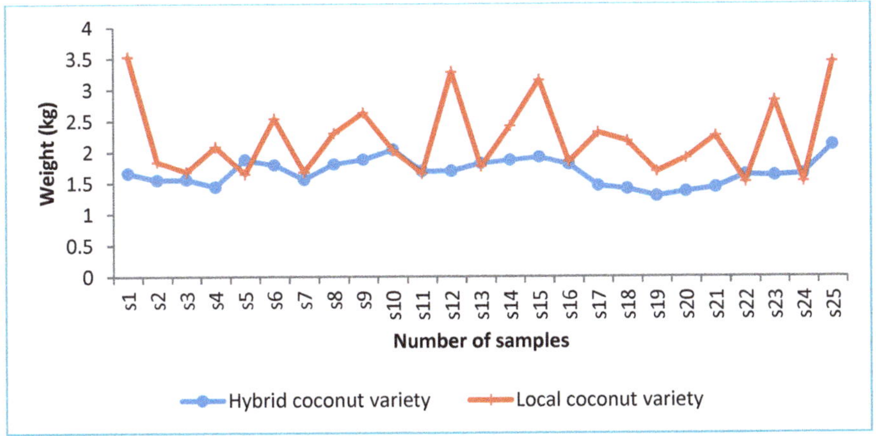


**Figure 6.** Plot of whole coconut waste samples and their weights.

### 3.3. Calorific Values of Charred and Uncharred Coconut Wastes

### 3.3.1. Uncharred Coconut Wastes (Husks and Shells)

The calorific value of the sun-dried coconut wastes (husks and shells) of both the local and hybrid varieties were analysed, and the results are presented in Table 2. The results indicated a mean calorific value of 11.54 ± 1.32 MJ/kg for the local coconut variety and 9.73 ± 0.33 MJ/kg for the hybrid variety. The calorific value, which is also known as heating value (q), is one of the important parameters that are considered when assessing a bioresource as a potential feedstock for fuel [20–23]. It is a measure of the amount of energy per unit mass or volume released on complete combustion. It is the amount of heat produced by the burning of 1 g of a substance and is measured in joules per gram (J/g).

**Table 2.** Calorific values of local and hybrid varieties of uncharred coconut waste (husk and shell).

| Readings | Mass of Sample (kg) | Calorific Value (MJ/kg) |
|---|---|---|
| Uncharred coconut waste of the local variety | | |
| 1 | 1.00 | 12.82 |
| 2 | 1.00 | 12.82 |
| 3 | 1.00 | 10.17 |
| | | 11.63 |
| Mean calorific value +/- Std. dev | | 11.54 ± 1.32 MJ/kg |
| Uncharred Coconut Waste of The Hybrid Variety | | |
| 1 | 1.00 | 9.394 |
| 2 | 1.00 | 9.762 |
| 3 | 1.00 | 10.044 |
| Mean calorific value +/- Std. dev | | 9.73 ± 0.33 MJ/kg |

The results indicate a variance in the calorific values obtained. The difference in calorific values is due to the chemical composition of the sample materials, in particular, the varying effect of lignin and extractive content [24]. The biomass of coconut is made up of cellulose, hemicellulose and lignin. There is about 65% cellulose in coconut shells, while lignin in coconut husk is almost 41% [25]. In addition to cellulose and lignin, coconut husk has pyroligneous acid, gas, charcoal, tar, tannin and potassium [26]. Further, with low amount of ash and more volatile matter, the husk make is appropriate for pyrolysis [26].

Amoako and Mensah-Amoah [27] determined the average calorific value of sun-dried uncharred coconut husks and shells to be 10.01 MJ/kg and 17.40 MJ/kg, respectively. These values are generally consistent with the results of 9.73 ± 0.33 MJ/kg to 11.54 ± 1.32 MJ/kg that were obtained in this study. The results also compare favourably with the calorific value of wood of 12–16 MJ/kg [24]. However, coconut waste burns fast, particularly the husk, and can therefore be used as fuel for less energy intense purposes, particularly for small-scale industrial heating, cooking and household applications [27]. Coconut husks and shells can therefore be attractive biomass fuels and are also a good source of charcoal [1,27].

### 3.3.2. Charred Coconut Wastes (Husks and Shells)

The calorific value of charred coconut wastes (husks and shells) was analysed and the results are presented in Table 3. The results indicated a mean calorific value of 21.307 ± 1.75 MJ/kg for charred coconut wastes of particle size P < 2 mm and 17.471 ± 5.53 MJ/kg for charred coconut wastes of P > 2 mm. From the results, the calorific value of the charred coconut waste is about 42% higher than the calorific value of the uncharred coconut waste. This is particularly significant for charred coconut wastes of particle size of P < 2 mm that are converted into briquettes for sustainable energy applications.

**Table 3.** Average calorific values of raw charred coconut waste (P < 2 mm, P > 2 mm).

| Samples | Average Calorific Value (MJ/kg) |
| --- | --- |
| Charred coconut waste (P < 2 mm) | 21.307 ± 1.75 |
| Charred coconut waste (P > 2 mm) | 17.471 ± 5.53 |

### 3.4. Moisture Content, Carbon Monoxide and Particulate Matter Emissions

Figure 7 shows the graph of moisture content and days of drying (sun drying) the coconut wastes. From the graphs, it is shown that as the drying days increased from 3 to 18 days, moisture content reduced as follows: day 3 (36.4%), day 6 (26.1%), day 9 (20.8%), day 12 (17.1%), day 15 (14.5%), and day 18 (10.3%). During drying some of the water in the waste material disappears and hence lessens the wet content. Under actual environmental settings, evaporation can happen because the actual wet content of the waste material is higher than its equilibrium moisture content, which is a factor of the material properties and environmental condition [28].

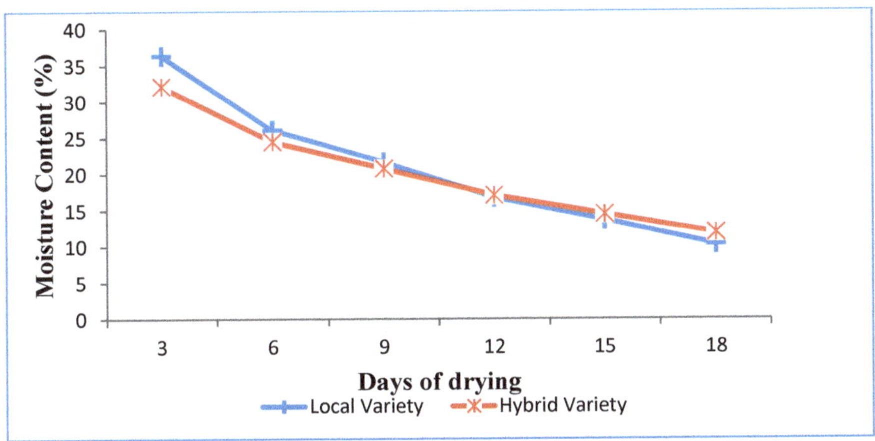


**Figure 7.** Moisture content of the coconut wastes by days of sun drying.

At harvest, moisture content of fresh coconut husks is around 29–35% [29]. In this study, it was observed that even after nearly one week (6 days) of open sun-drying, the moisture content of the coconut wastes reduced marginally to about 26%. Huda et al. [30] reported that high moisture content of biomass results in poor ignition and reduces the combustion temperature, which in turn affects the combustion of the products and quality of combustion. In general, the moisture content of biomass resources, especially wood, changes the calorific value of the latter by lowering it [31]. The explanation is that part of the energy released during the combustion process is spent in water evaporation. According to Raghavan [28], dry coconut husks with a moisture content of 10% had been used as fuel for the drying of copra in an island community in the Philippines.

Since local food vendors use either the raw unprocessed coconut wastes or dry them in the open sun for a number of days, it was essential to study the moisture content over time in order to understand its effects when coconut wastes are utilized as fuel for domestic cooking in fuel-constraint and energy-poor communities. Properly seasoned firewood has a moisture content below 20% [30]. Now, if we assume this measure for coconut biomass, then we can infer that to use properly seasoned coconut husks and shells with moisture content below 20%, it is likely to take 9 to 18 sun drying days to achieve moisture contents of 10% to 20%.

Figure 8 shows the graphs of carbon monoxide (CO) and moisture content of the hybrid, and local coconut wastes. CO concentrations for coconut wastes sun-dried for 3 days were 7.9 ppm for the local coconut wastes and 12.1 ppm for the hybrid coconut wastes. The values reduced from 7.9 to 7.1 ppm and 12.1 to 10.1 ppm over the 3–18 drying days, resulting in a steady reduction in CO concentration of 10–17% for the studied varieties. The CO concentrations generally decreased with decreasing moisture content over the drying period for both varieties. The smoke produced when burning the husks and shells that were being dried gradually changed from thick white to light smoke in 3 to 15 days. This is an indication of the fact that there was a decreasing amount of volatile gases including water vapour that resulted in the change in the concentration and colour of the smoke (see Figure 5).

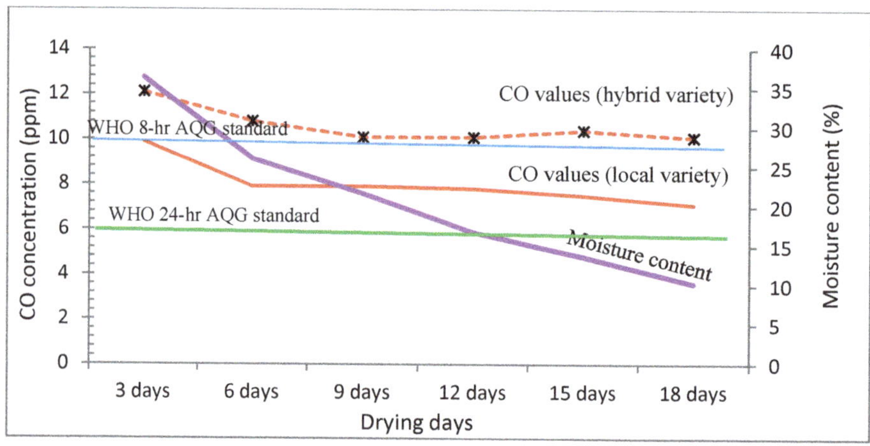


**Figure 8.** Graphs of CO and moisture content of the hybrid and local coconut wastes.

From the graphs in Figure 8, the CO emissions measured were higher than the World Health Organisation (WHO) indoor air quality guideline (AQG) values at 25 °C, 1 atmosphere. Some suggested tips that are applicable to typical indoor exposure are as follows: 10 mg/m$^3$ (8.73 ppm) for 8 h (average concentration, low to moderate exercise); and 7 mg/m$^3$ (6.11 ppm) for 24 h (average concentration, with the assumption that during the exposure people are not sleeping and alert without doing any exercise [32]. However, according to US EPA, outdoor maximum levels should be 35 ppm (1 h averaging) and 9 ppm (8 h averaging), while WHO limits CO concentrations of 90 ppm to 10 ppm based on exposure time of 15 min to 8 h respectively [33].

Overall, the observation is that the burning of fresh unprocessed biomass materials with high moisture levels such as fresh coconut wastes that are used for domestic cooking and other applications are likely to produce higher concentrations of carbon monoxide than charred biomass materials with relatively low moisture content. High moisture levels of fresh biomass materials do not only result in high CO emissions, but also affect the calorific value of the materials.

After sun-drying the coconut waste from 3 to 18 days, data on particulate matter (PM$_{2.5}$) emissions measured are presented in Figure 9. From the graphs, the hybrid coconut wastes showed higher PM$_{2.5}$ values of min = 994 ug/m$^3$ and max = 1 425 ug/m$^3$ than the local coconut wastes with PM$_{2.5}$ values of min = 933 ug/m$^3$ and max = 1169 ug/m$^3$. Comparing the values to [34,35] air quality guidelines (AQG) of PM$_{2.5}$ = 10 µg/m$^3$ annual mean and 25µg/m$^3$ 24 h mean, it is obvious that the PM$_{2.5}$ from study results were relatively high. The implication is that people who use raw unprocessed waste coconut husks and shells as fuel are under the risks of the adverse effects of PM$_{2.5}$ emissions as a result of the combustion method and conditions. Household combustion methods of biomass are of low energy conversion efficiency and therefore result in high pollutant emissions [36].

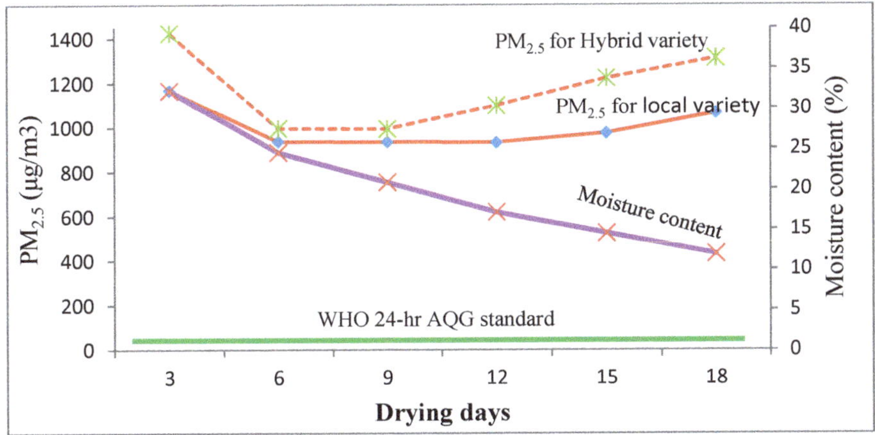


**Figure 9.** Graphs of PM and moisture content of the hybrid and local coconut wastes.

## 4. Conclusions and Recommendations

This study assessed the waste proportions, caloric values, and pollutant emissions from the burning of raw uncharred and charred coconut wastes in Ghana.

The results indicate that 62–65% of the whole coconut fruit can be generated as wastes in the form of husks and shells. This amount constitutes a potential bioenergy resource that can be considered as an alternative to firewood and hence can be used as fuel for small-scale electricity production, industrial heating, cooking and household applications.

In this study, the calorific values of the raw uncharred and charred coconut wastes were determined. The average calorific value of the charred coconut wastes was 42% more than that of the uncharred coconut wastes. The moisture content of the raw uncharred coconut wastes might have influenced the relatively low calorific value. The implication is that with relatively high calorific value, charred coconut wastes can be considered to be a better fuel than the raw uncharred coconut wastes that are being burned as domestic fuel, particularly in energy-poor households.

With regard to smoke emissions, the study found that as water evaporated gradually from the raw uncharred coconut wastes during the combustion process, CO emissions generally decreased to a level considered to be within the WHO AQG for 8 h, even though it was above the WHO AQG for 24 h. However, PM$_{2.5}$ pollutant emissions did not meet the WHO 24 h indoor air quality guidelines value at 25 °C, 1 atm. This suggests that charred coconut wastes would likely produce less CO pollutant emissions than the raw uncharred coconut wastes. To effectively utilise coconut wastes as a bioenergy resource for biochar briquette fuel, there is the need to produce biochar for briquettes in a controlled system to maximise the calorific value and minimise smoke emissions.

**Author Contributions:** Conceptualization, G.Y.O., D.Y.A. and E.M.; Survey, experiments, and data gathering, D.Y.A., G.Y.O. and E.M.; Scientific content and structure—G.Y.O., D.Y.A., E.M. and R.O. Literature review—D.Y.A., G.Y.O. and R.O. Initial draft review—G.Y.O., D.Y.A., R.O., C.K.K.S. and E.A.A. Second draft review and content fine-tuning—G.Y.O., R.O., C.K.K.S., E.A.A. and E.M. All authors have read and agreed to the published version of the manuscript.

**Funding:** This research received no external funding.

**Acknowledgments:** Laboratory equipment used for testing and measurements was financed by UNDP Ghana and supported by the USAID/MIT D-Lab International Development Innovation Network (IDIN) programme. The authors are thankful to the staff of the Cookstove Testing and Expertise Laboratory (C-Lab) of the Technology Consultancy Centre, KNUST for the experimental setup, testing, and data generation that made it possible for us to write the paper.

**Conflicts of Interest:** The authors declare no conflict of interest.

## References


1. UNEP. *Technologies for Converting Waste Agricultural Biomass to Energy*; UNEP–United Nations Environment Programme: Nairobi, Kenya; Division of Technology, Industry and Economics International Environmental Technology Centre Osaka: Osaka, Japan, 2013; pp. 1–214.
2. UNCTAD. National Green Export Review of Vanuatu: Copra-Coconut, Cocoa-Chocolate and Sandalwood, United Nations Conf. Trade Dev. (UNCTAD), 2016, United Nations Publ. Available online: https//unctad.org/en/PublicationsLibrary/ditcted2016d1_en.pdf (accessed on 5 May 2019).
3. Rahamat, S.F.; Manan, W.N.H.W.A.; Jalaludin, A.A.; Abllah, Z. Enamel subsurface remineralization potential of virgin coconut oil, coconut milk and coconut water. *Mater. Today Proc.* **2019**, *16*, 2238–2244. [CrossRef]
4. Lu, X.; Su, H.; Guo, J.; Tu, J.; Lei, Y.; Zeng, S.; Chen, Y.; Miao, S.; Zheng, B. Rheological properties and structural features of coconut milk emulsions stabilized with maize kernels and starch. *Food Hydrocoll.* **2019**, *96*, 385–395. [CrossRef]
5. de Oliveira, E.; Quitete, F.T.; Bernardino, D.N.; Guarda, D.S.; Caramez, F.A.H. Maternal coconut oil intake on lactation programmes for endocannabinoid system dysfunction in adult offspring. *Food Chem. Toxicol.* **2019**, *130*, 12–21. [CrossRef] [PubMed]
6. Akpro, L.A. Phytochemical compounds, antioxidant activity and non-enzymatic browning of sugars extracted from the water of immature coconut (Cocos nucifera L.). *Sci. Afr.* **2019**, *6*, e00123. [CrossRef]
7. Ding, K. A rapid and efficient hydrothermal conversion of coconut husk into formic acid and acetic acid. *Process Biochem.* **2018**, *68*, 131–135. [CrossRef]
8. Anuar, M.F.; Fen, Y.W.; Zaid, M.H.M.; Matori, K.A.; Khaidir, R.E.M. Synthesis and structural properties of coconut husk as potential silica source. *Results Phys.* **2018**, *11*, 1–4. [CrossRef]
9. Talat, M.; Mohan, S.; Dixit, V.; Singh, D.K.; Hasan, S.H.; Srivastava, O.N. Effective removal of fluoride from water by coconut husk activated carbon in fixed bed column: Experimental and breakthrough curves analysis. *Groundw. Sustain. Dev.* **2018**, *7*, 48–55. [CrossRef]
10. Muharja, M.; Junianti, F.; Ranggina, D.; Nurtono, T.; Widjaja, A. An integrated green process: Subcritical water, enzymatic hydrolysis, and fermentation, for biohydrogen production from coconut husk. *Bioresour. Technol.* **2018**, *249*, 268–275. [CrossRef]
11. Buamard, N.; Benjakul, S. Effect of ethanolic coconut husk extract and pre-emulsification on properties and stability of surimi gel fortified with seabass oil during refrigerated storage. *LWT Food Sci. Technol.* **2019**, *108*, 160–167. [CrossRef]
12. Ram, M.; Mondal, M.K. Comparative study of native and impregnated coconut husk with pulp and paper industry waste water for fuel gas production. *Energy* **2018**, *156*, 122–131. [CrossRef]
13. Narayanankutty, A.; Illam, S.P.; Raghavamenon, A.C. Health impacts of different edible oils prepared from coconut (Cocos nucifera): A comprehensive review. *Trends Food Sci. Technol.* **2018**, *80*, 1–7. [CrossRef]
14. Talha, N.S.; Sulaiman, S. In situ transesterification of solid coconut waste in a packed bed reactor with CaO/PVA catalyst. *Waste Manag.* **2018**, *78*, 929–937. [CrossRef]
15. GSS. *Main Report. Ghana Living Standard Survey Round 6 (GLSS 6)*; Ghana Statistical Service (GSS): Accra, Ghana, 2014.
16. Lee, K.; Park, E. Residential air quality in wood burning houses in Costa Rica. *Proc. Indoor Air* **2002**, *4*, 612–617.
17. Peters, A.; Wichmann, H.E.; Tuch, T.; Heinrich, J.; Heyder, J. Respiratory effects are associated with the number of ultrafine particles. *Am. J. Respir. Crit. Care Med.* **1997**, *155*, 1376–1383. [CrossRef] [PubMed]
18. Kelly, M.; Donnelly, R. *The Humongous Book of Statistics Problems*; Penguin Group: New York, NY, USA, 2009; ISBN 978-1-59257-865-8.
19. Zafar, S. Energy Potential of Coconut Biomass, BioEnergy Consult. Available online: https//www.bioenergyconsult.com/coconut-biomass/ (accessed on 17 October 2019).
20. ÖzyuğUran, A.; Yaman, S. Prediction of calorific value of biomass from proximate analysis. *Energy Procedia* **2017**, *107*, 130–136. [CrossRef]
21. Lu, Z. Feasibility study of gross calorific value, carbon content, volatile matter content and ash content of solid biomass fuel using laser-induced breakdown spectroscopy. *Fuel* **2019**, *258*, 116150. [CrossRef]
22. Tang, J.P.; Lam, H.L.; Aziz, M.K.A.; Morad, N.A. Enhanced biomass characteristics index in palm biomass calorific value estimation. *Appl. Therm. Eng.* **2016**, *105*, 941–949. [CrossRef]

23. Ozyuguran, A.; Akturk, A.; Yaman, S. Optimal use of condensed parameters of ultimate analysis to predict the calorific value of biomass. *Fuel* **2018**, *214*, 640–646. [CrossRef]
24. Kaltschmitt, M.; Hartmann, H.; Hofbauer, H. *Energie Aus Biomasse. Grundlagen, Techniken und Verfahren*, 2nd ed.; Springer: Berlin, Germany, 2009.
25. Wang, Q.; Sarkar, J. Pyrolysis behaviors of waste coconut shell and husk biomasses. *Int. J. Energy Prod. Mgmt.* **2018**, *3*, 34–43. [CrossRef]
26. Zafar, S. Coconut Husk—Energy Potential of Coconut Biomass. BioEnergy Consult March 15. 2019. Available online: https://www.bioenergyconsult.com/tag/coconut-husk/ (accessed on 20 January 2020).
27. Amoako, G.; Mensah-Amoah, P. Determination of calorific values of coconut shells and coconut husks. *J. Mater. Sci. Res. Rev.* **2018**, *2*, 1–7.
28. Raghavan, K. Biofuels from Coconut. FACT, August 2010, pp. 1–107. Available online: https://energypedia.info/images/f/f9/EN-Biofuels_from_Coconuts-Krishna_Raghavan.pdf (accessed on 25 September 2019).
29. Tooy, D.; Nelwan, L.; Pangkerego, F. Evaluation of biomass gasification using coconut husks in producing energy to generate small-scale electricity. In Proceedings of the International Conference on Artificial Intelligence, Energy and Manufacturing Engineering (ICAEME'2014), Kuala Lumpur, Malaysia, 9–10 June 2014; pp. 84–88. [CrossRef]
30. Huda, N.; Rashid, M.; Hasfalina, C. Particulate emission from agricultural waste fired boiler. *Int. J. Innov. Appl. Stud.* **2014**, *8*, 1265–1295.
31. Krajnc, N. *Woodfuels Handbook*; Food Agricultural Organization (FAO) of the United Nations: Pristina, Kosovo, 2015; ISBN 978-92-5-108728-2. Available online: https://roycestreeservice.com/wp-content/uploads/Wood-Fuels-Handbook.pdf (accessed on 20 January 2020).
32. WHO. *WHO Guidelines for Indoor Air Quality: Selected Pollutants*; WHO Reg. Off. Eur.: Copenhagen, Denmark, 2010; ISBN 978928902134.
33. WHO. *Carbon Monoxide: Air Quality Guidelines. Chapter 5.5*, 2nd ed.; Regional Office for Europe: Copenhagen, Denmark, 2000; Available online: http://www.euro.who.int/__data/assets/pdf_file/0020/123059/AQG2ndEd_5_5carbonmonoxide.PDF (accessed on 20 January 2020).
34. Adam, T. *W179 Wood Products Information—Moisture Content of 'Seasoned' Firewood*; UT Extension Publications, The University of Tennessee Agricultural Extension Service: Knoxville, TN, USA, 2010.
35. WHO. *WHO Indoor Air Quality Guidelines: Household Fuel Combustion*; Publ. World Heal. Organ. (WHO): Geneva, Switzerland, 2014; ISBN 9789241548878 (print), ISBN 9789241548885 (CD-ROM).
36. Chen, J.; Li, C.; Ristovski, Z.; Milic, A.; Gu, Y.; Wang, S.; Hao, J.; Zhang, H.; He, C.; Guo, H.; et al. A review of biomass burning: Emissions and impacts on air quality, health and climate in China. *Sci. Total Environ.* **2017**, *579*, 1000–1034. [CrossRef] [PubMed]

*Article*

# Different Pyrolysis Process Conditions of South Asian Waste Coconut Shell and Characterization of Gas, Bio-Char, and Bio-Oil

**Jayanto Kumar Sarkar and Qingyue Wang ***

Graduate School of Science and Engineering, Saitama University, 255 Shimo-okubo, Sakura-ku, Saitama 338-8570, Japan; joysarkaric@gmail.com
* Correspondence: seiyo@mail.saitama-u.ac.jp; Tel.: +81-(48)-8583733

Received: 3 March 2020; Accepted: 13 April 2020; Published: 16 April 2020

**Abstract:** In the present study, a series of laboratory experiments were conducted to examine the impact of pyrolysis temperature on the outcome yields of waste coconut shells in a fixed bed reactor under varying conditions of pyrolysis temperature, from 400 to 800 °C. The temperature was increased at a stable heating rate of about 10 °C/min, while keeping the sweeping gas (Ar) flow rate constant at about 100 mL/min. The bio-oil was described by Fourier transform infrared spectroscopy (FTIR) investigations and demonstrated to be an exceptionally oxygenated complex mixture. The resulting bio-chars were characterized by elemental analysis and scanning electron microscopy (SEM). The output of bio-char was diminished pointedly, from 33.6% to 28.6%, when the pyrolysis temperature ranged from 400 to 600 °C, respectively. In addition, the bio-chars were carbonized with the expansion of the pyrolysis temperature. Moreover, the remaining bio-char carbons were improved under a stable structure. Experimental results showed that the highest bio-oil yield was acquired at 600 °C, at about 48.7%. The production of gas increased from 15.4 to 18.3 wt.% as the temperature increased from 400 to 800 °C. Additionally, it was observed that temperature played a vital role on the product yield, as well as having a vital effect on the characteristics of waste coconut shell slow-pyrolysis.

**Keywords:** fixed bed; pyrolysis yield; temperature; coconut shell; characterization; SEM

---

## 1. Introduction

Energy is important for agricultural production, electrical generation, transportation and industrial progress, and other economic sectors [1]. Fossil fuel is the leading source of energy and is interred deep inside the Earth. However, these resources are insufficient and are not capable of fulfilling for long the growing global energy requirement [2]. Moreover, there are adverse influences from the consumption of fossil fuels on climate, atmospheric pollution, acid rain, and global warming, for example [3]. For this reason, alternative and durable energy sources are essential to fulfill this rising demand for energy. With the swift expansion in its overall energy application, and the sustainability and expanding environmental impact from fossil fuels, biomass fuels, as sustainable energy sources, have progressively been considered a key choice to replace traditional fossil fuels. Lignocellulosic biomass residues are byproducts or the waste from processed agricultural products, which is an enormous and inexpensive source of sustainable energy that does not affect food or feed supplies. It is outstandingly inexpensive, in comparison to conventional fossil fuels, based on energy supply. Currently, biomass and residues account for 10% to 15% of the world's energy demand [4].

The pyrolysis of biomass has received increasing attention. Within the previous decade, pyrolysis has been the most encouraging thermochemical technique to provide energy from biomass. Pyrolysis is a procedure in which the thermal deterioration of the organic elements of biomass is formed and maintains the environmental inanimate responses in order to acquire energy. The pyrolysis of the biomass brings about three items: bio-oil, gas, and bio-char. Bio-oil is created in the pyrolysis procedure and has a prosperity as feedstock for electricity generation because it contains a huge amount of energy that is practically identical to the petroleum products after upgradation [5]. Non-condensable gases consist of $CH_4$ (methane), $H_2$ (hydrogen), CO (carbon monoxide), and $CO_2$ (carbon dioxide), which might be burned for energy recuperation or for the creation of syngas. Bio-char is a hard, carbon-rich element that is thermally durable for biomass or some other organic elements [6]. Additionally, bio-char created during pyrolysis contains a large amount of energy, which at times is equivalent to the coal utilized as fuel in ventures [7,8]. The microporous formation of bio-char and its large amount of carbon content makes it valuable for a few modern applications. Moreover, to promote the efficiency of soil, bio-char can be effectively employed from an agricultural perspective. The application of bio-char within soil builds the pace of carbon sequestration in soil. It hinders the pace of supplement deterioration in soil, and thus, improves soil quality, such as the fertility of soil [9–12]. In electricity generation, the large amount of carbon content suggests that bio-char can be utilized as a fuel.

Coconut shell accounts for an ordinary biomass waste, which is in enormous proportions in all the tropical regions in Asia, Africa, and Latin America. Coconut shell is an inexpensive resource because it is found in over 90 nations around the world [13]. The coconut shell is produced from oil manufacturing, several agro industrial activities, and different utilizations of coconut. This coconut shell byproduct waste needs to be recycled. It can be a significant source of energy if it is properly utilized. Numerous investigators have explored the pyrolysis of other biomasses, for example, sawdust, straw, mangaba seed, corncob, miscanthus, olivekernel, almond shell, and regnum stalks [14–17]. However, the studies related to coconut shell biomass are outnumbered. Raveendran et al. [18] investigated thirteen biomass samples, in which coconut shell was one. However, insignificant attention was given to these materials. Hoque and Battacharya [19] only studied the coconut shell gasification item from fluidized and spouted bed gasifiers, at a temperature range from 607 to 842 °C. Tritti et al. [20] explored the product characteristic of the fast pyrolysis of coconut shell biomass using FTIR (Fourier transform infrared spectroscopy). Tsamba et al. [21] only studied pyrolysis characteristics and the global kinetics of coconut and cashew nut shells using a thermogravimetric analyzer. Solid, liquid, and gaseous productions, acquired by coconut shell flash and fast pyrolysis, were formerly described [20,22,23]. Nevertheless, comprehensive investigations of the pyrolysis yield and utilization of the byproduct are rare, especially for the effect of the reaction condition and characterization. Thus, the objectives of this paper were to evaluate the effect of pyrolysis temperature on the product yields, as well as to clarify the characterization and the parceling of the mass of coconut shell waste pyrolysis products under varying conditions of pyrolysis temperatures.

## 2. Materials and Methods

### 2.1. Sample Preparation

Mature and properly ripened brown coconuts (*Cocos nucifera*) were collected from the local fruit market in Jessore, located in Khulna division, Bangladesh. The coconut shells were segregated from the copra and husk. The coconut shells were dried in the sunlight for several days in order to withdraw the extract. The samples were crushed and sieved to less than 250 μm by a crusher and sieve shaker, dried at a natural temperature, and then exposed for one hour in the laboratory atmosphere in order to obtain the air-dried samples. Finally, the samples were placed in the dissector. Proximate and ultimate analyses were performed on air-dried samples. The ultimate or elemental analysis was executed by employing a carbon, hydrogen and nitrogen coder (CHN) (MT-5 Yanaco, Co. Ltd., Japan). Moisture, ash, and volatile matter were found by adopting the Japanese Industrial Standard Code procedures. Fixed carbon (FC) was estimated as follows:

$$FC\ (wt.\%) = 100 - \{Moisture(wt.\%) + Ash(wt.\%) + Volatile\ Matter(wt.\%)\} \tag{1}$$

## 2.2. Experimental Apparatus for Fixed Bed Pyrolysis

Figure 1 illustrates the different experimental arrangements for biomass pyrolysis and gasification. The main components include a pyrolysis system, a trapping system for condensable products, a gas feeding system, a system for the measurement of gaseous products, and a liquid decomposition system. Two stainless steel coupling reactors, each with an inner diameter dimension of 21.4 mm and a length of 500 mm, were employed in the liquid decomposition system and the pyrolysis system. Biomass was positioned in the 40 μm mesh of each reactor. Furthermore, two distinct electric furnaces governed the independent heating processes of the apparatuses. Additionally, the apparatuses could play a significant role in the heterogeneous reaction amid liquid and ash. The paired stainless steel pipes between the liquid tapping system and the reactors were heated to 300–400 °C in order to avoid contraction of the liquid. Test tubes were inserted in order to collect the condensable products and glass beads were used to boost the collection capability. A cooling bath temperature of –3 °C was maintained with the application of water and ice. Gas chromatography was used to calibrate the different gaseous products.

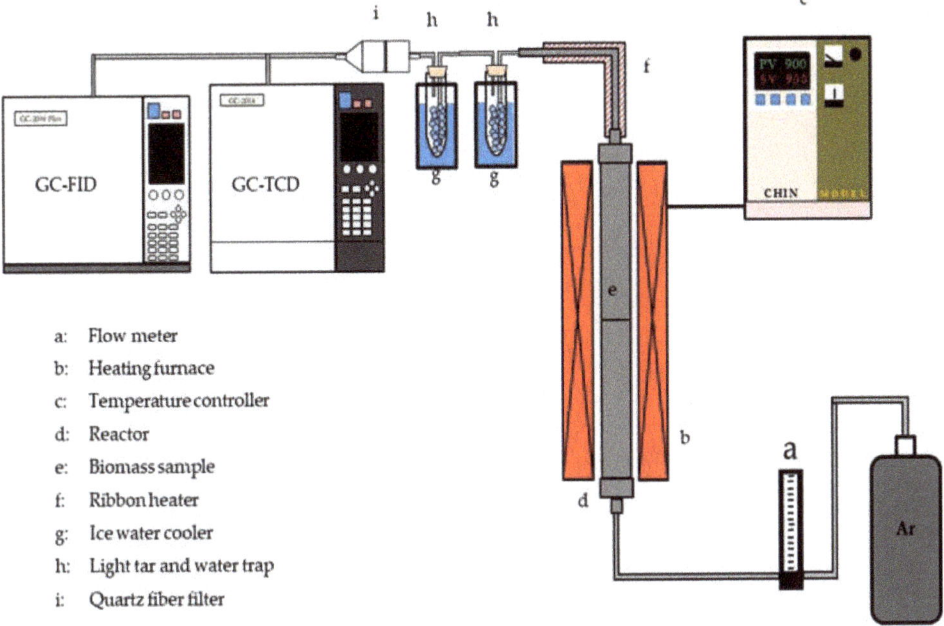


a:  Flow meter
b:  Heating furnace
c:  Temperature controller
d:  Reactor
e:  Biomass sample
f:  Ribbon heater
g:  Ice water cooler
h:  Light tar and water trap
i:  Quartz fiber filter

**Figure 1.** Experimental apparatus for pyrolysis and gasification product measurements.

## 2.3. Experimental Procedure

In this experiment, approximately 4 g of the dried sample was positioned on a mesh portion inside the reaction tube. A conduit was connected and installed in an electric furnace. Assembled traps and measure weights were then connected with the liquid trap and filter holder. The joint and upper part of the reaction tube was warped with a ribbon heater. A flow of Ar was supplied in order to create an appropriate reductive atmosphere within the reactor. Thereafter, the temperature control program was set at a heating rate of 10 °C/min, from room temperature (about 25 °C), to 400–800 °C, and it started heating. After the experiment was completed, the electric furnace was opened and the reaction tube was cooled for 1 h using an electric fan. The gaseous products ($H_2$, CO, $CO_2$, $CH_4$) were evaluated based on the measurement results by gas chromatography (GC-2010, Shimadzu Co. Ltd., Japan). The bio-oil (light-weight tar) produced was found by the weight difference method of the liquid entrapment before and after the experiment. However, the char was defined as the residual product inside the reactor after performing the experiment. Heavy tar was condensed at an ambient temperature, which caused clogging in the pyrolysis reactor and pipes, and was calculated as follows:

$$\text{Heavy tar} = \text{Biomass} - (\text{Bio-oil} + \text{char} + \text{total gas}) \tag{2}$$

*2.4. Measurement of Waste Coconut Shell Pyrolysis Product*

The ultimate or elemental analyses of solid bio-char were performed by employing a CHN corder (MT-5 Yanaco, Co. Ltd., Japan). According to Dulong's formula, higher heating values (HHV) were estimated as follows:

$$\text{Heating value (MJ/kg)} = 0.338\,C + 1.428\,(H - O/8) + 0.095\,S \tag{3}$$

where C, H, O, and S are the carbon, hydrogen, oxygen, and sulfur, respectively. These were the fundamental compositions in the material weight percentages.

Surface morphologies of bio-char were visualized by the scanning process of a scanning electron microscope (SEM, SU1510) at a fixed voltage of 15 kV.

The functional groups of coconut shell sample pyrolysis were quantified by Fourier transform infrared spectroscopy (FTIR). The pyrolysis liquid samples were covered by thin plates of potassium bromide (KBr). Additionally, after placement of the pellet within the FTIR instrument, the functional groups of liquid samples were evaluated. The infrared spectrum was documented within a range of 500 to 4000 $cm^{-1}$.

*2.5. Statistical Analysis*

We conducted one-way analysis of variance (ANOVA)s to test the effect of temperature on all the measured parameters, namely the product yield (bio-char, bio-oil, and gas), product content (H, C, N, and O), HHV, and yields of distinct gases, i.e., CO, $CO_2$, $CH_4$, and $H_2$, followed by a post-hoc multiple comparison test with least significant difference (LSD). A square-root function was used to transfer the data in order to maintain the homogeneity of the variances of these analyses, and Levene's test of equality was applied to check the homogeneity of the variance of the data that were used. After that, $p \leq 0.05$ was considered to be significant for all experimental data analyses by using the statistical software IBM SPSS statistics 24.0 (whereas $p \leq 0.05$ means that the test hypothesis was statistically significant).

## 3. Results and Discussion

*3.1. Characterization of Raw Biomass*

The ultimate and proximate analysis and HHV for the waste coconut shell is displayed in Table 1. From the ultimate analysis results of waste coconut shell, the carbon (C), hydrogen (H), nitrogen (N), and oxygen (O) amounts were 39.22, 4.46, 0.02, and 56.10 wt.%, respectively. The HHV of waste coconut shell was found to be 9.62 MJ/kg. However, the mass amount of oxygen content may convey adverse effects on the HHV. Low nitrogen substance is crucial, due to the fact that higher N percentages might result in toxic $NO_2$ emissions throughout pyrolysis [24]. From the proximate analysis results of waste coconut shell, the values of moisture content, volatile matter, fixed carbon, and ash were 7.82, 79.91, 12.04, and 0.23 wt.%, respectively. A quantitative comparison with previous studies related to coconut shell is presented in Table 2. In Table 2, all the elemental and proximate analyses were found to be different when compared with the present study, due to both the elemental and proximate components being significantly dependent on the maturity of coconut, soil quality, and environmental condition during cultivation. Waste coconut shell has a high volatile content, which is good for the pyrolysis process. A low ash amount is significant because a high ash amount can trigger aggregation in experimental procedures and can result in unproductive heat transfer rates. Low combustion, spillover processing costs, difficulties in the disposal, and waned energy conversion are potential reasons for an undesirable amount of ash [25,26]. Inorganic minerals present in ash greatly affect biomass pyrolysis mechanisms [27].

**Table 1.** Proximate and ultimate analysis of waste coconut shell.

| Ultimate Analysis (wt.%) | | Proximate Analysis (wt.%) | |
|---|---|---|---|
| Carbon | $39.22 \pm 0.71$ | Moisture content | $7.82 \pm 0.02$ |
| Hydrogen | $4.46 \pm 0.08$ | Volatile matter | $79.91 \pm 0.05$ |
| Nitrogen | $0.22 \pm 0.02$ | Fixed carbon | $12.04 \pm 0.04$ |
| Oxygen | $56.10 \pm 0.81$ | Ash | $0.23 \pm 0.003$ |
| Sulfur | ND | | |
| HHV (MJ/kg) | $9.62 \pm 0.50$ | | |

Note: ND: Not detected, HHV: higher heating values.

**Table 2.** Comparison of proximate and ultimate analysis of coconut shell.

| Author | Proximate and Ultimate Analysis of Coconut Shell | | | | | | | | | Reference |
|---|---|---|---|---|---|---|---|---|---|---|
| | Elemental Analysis (wt.%) | | | | | Proximate Analysis (wt.%) | | | | |
| | C | H | N | O | S | Moisture | Volatile | Fixed Carbon | Ash | |
| Rout et al. | 64.23 | 6.89 | 0.77 | 27.61 | 0.50 | 10.1 | 75.5 | 11.2 | 3.2 | [13] |
| Sundaram et al. | 53.73 | 6.15 | 0.86 | 38.45 | 0.02 | | 72.93 | 19.48 | 0.61 | [28] |
| Tsai et al. | 63.45 | 6.73 | 0.43 | 28.27 | 0.17 | 11.26 | 79.59 | | 3.38 | [23] |
| Tsamba et al. | 53.9 | 5.7 | 0.1 | 39.44 | 0.02 | | 74.9 | 24.4 | 0.7 | [21] |

## 3.2. Product Yields under Operating Variables

Coconut shell biomass was pyrolyzed using a fixed bed reactor under several pyrolysis temperatures. Each experiment was repeated at least three times. Figure 2 represents the product yields under several pyrolysis temperatures. The product yields varied significantly across the temperature ranges (bio-char: $F_{4,10} = 1107.48$, $p \leq 0.0001$; bio-oil: $F_{4,10} = 70.54$, $p \leq 0.0001$; gas: $F_{4,10} = 207.65$, $p \leq 0.0001$; and heavy tar: $F_{4,10} = 62.31$, $p \leq 0.0001$) (Figure 2). The bio-char at 400 to 600 °C was significantly higher than 700 and 800 °C ($p \leq 0.0001$), and they were significantly different from each other ($p \leq 0.0001$). However, no significant differences ($p = 0.251$) were observed for the bio-char in the temperature range from 700 to 800 °C. The proportion of bio-char relative to the entire number of biomass samples was reduced from 33.6 to 27.6 wt.% for coconut shell biomass when the pyrolysis temperature was raised from 400 to 700 °C. With the continually rising temperature, no momentous difference in the solid bio-char yields was observed. The pyrolysis of biomass samples at the central part intensified with the rise in temperature, which was one of the probable causes, and other researchers were satisfied with these outputs [29]. When the pyrolysis temperature rose from 400 to 600 °C, the product properties of bio-oil also increased from 47.5 to 48.7 wt.%, respectively. In addition, with the further rise in the pyrolysis temperature, from 600 to 800 °C, the product properties of bio-oil declined from 48.7 to 46.7 wt.%. Pointedly, the weight yield of bio-oil acquired its maximum value at 600 °C under all the considered cases. However, this phenomenon might have caused a significant impact on the secondary cracking behaviors of the volatiles at advanced pyrolysis temperatures (500 to 600 °C). Corresponding outcomes were accessed from the other research results [30,31]. Table 3 shows a comparison of the product yields, namely char, liquid, and gas, which are more or less similar to our results. Pyrolysis temperature was found to have an important effect of non-condensable gas yields. With the further expansion of the gas products, there was a more significant effect at higher temperatures, mainly because of the secondary cracking attitudes of the volatiles. Additionally, throughout the secondary decomposition process, the bio-char may generate non-condensable elements, which would boost gas yields in increments of different pyrolysis temperatures.

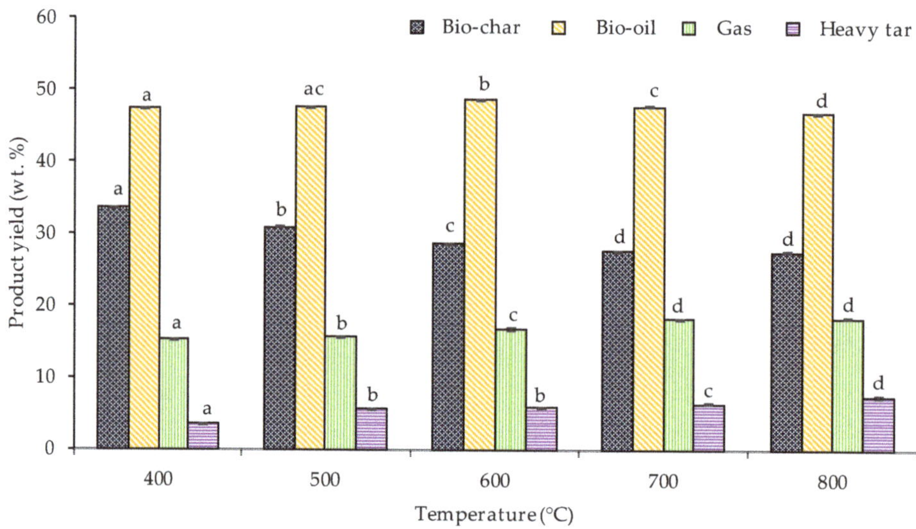


**Figure 2.** Product yields of pyrolysis products against varying conditions of temperature, where each bar indicates the mean ± standard error [*SE* (*n* = 3)]. Note: distinct letters in each specific yield bar indicate significant differences ($p \leq 0.05$).

**Table 3.** Comparison of pyrolysis product yield of coconut shell.

| Reactor | Pyrolysis Type | Temperature (°C) | Findings | Ref. |
|---|---|---|---|---|
| Semi-bath | Slow | 450–600 °C | Maximum yield of liquid was found to be 49.5 wt.% at 575 °C, whereas gas yield decreased from 29 to 24 wt.%. In addition, char yield followed a decreasing trend from approximately 32% to 25.4% when the temperature increased from 400 to 600 °C. | [13] |
| Fixed bed | Slow | 400–600 °C | The yield of liquid and gaseous products increased from 38 to 43 wt.% and 30 to 33 wt.% at the temperature of 400–600 °C, whereas the char yield decreased from 32 to 22 wt.%. | [28] |
| Induction heating | First | 400–800 °C | A significant increased trend on the yield of liquid products was observed when temperature increased from 400 to 500 °C, whereas an opposite trend was observed for char yield. | [23] |
| Fixed bed | Slow | 400–800 °C | Bio-oil product properties increased until the temperature reached 600 °C, however, with the further increase of temperature, it followed a decreased trend. These results were consistent with Sundaram and Natarajan et al. [28]. Non-condensable gas significantly increased from 15.37 to 18.34 wt.% for the temperature range considered. However, the proportion of char yield followed a decreased trend from 33.6 to 27.6 wt.% when the temperature varied from 400 to 700 °C. Similar outcomes were observed by Sundaram and Natarajan et al. [24]. | This study |

### 3.3. Characterization of Waste Coconut Shell Bio-Chars

#### 3.3.1. Elemental Analysis of Bio-Char

The fundamental components of the bio-char underwent a widespread shift on account of the dejection of maximum volatiles during the pyrolysis procedure. Figure 3 shows the fundamental components of bio-char at various pyrolysis temperatures. Elements C and H in the bio-char differed significantly across the temperature ranges (C: $F_{4,10} = 9.93$, $p = 0.002$; H: $F_{4,10} = 1716.56$, $p \leq 0.0001$; and O: $F_{4,10} = 62.31$, $p = 0.006$), except N ($F_{4,10} = 3.001$, $p = 0.072$) (Figure 3). The major elements were C, O, and a small amount of H and N. With the increment of pyrolysis temperature, an increasing trend in the carbon components of the bio-char from coconut shell was observed. Under 400 °C, the components of O and H were reduced, due to the condensation and withdrawal of the OH bonds [32]. With rising temperatures, higher volatiles were also released, which lead to a reduction in O and H components. A reduction in O elements was shown when the temperature was over 400 °C; however, the reduced range was small. It implies that the oxygen content involved in the practical units of the fission process was evaporated in the lower temperature zone [33]. The N element exhibited a slow ascending trend with the increment in temperature. The reduction in nitrogen content may be due to the cleavage of nitrogen-containing practical units and the discharge of different gaseous products that contain basic N contents [34].

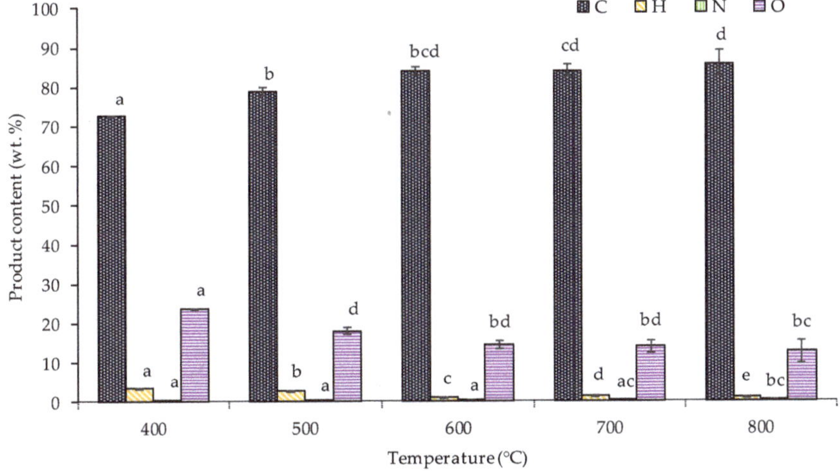


**Figure 3.** Effect of temperature on fundamental content of coconut shell bio-char against varying conditions of temperature, where each bar indicates the mean ± standard error [*SE* (*n* = 3)]. Note: distinct letters in each specific yield bar indicate significant differences ($p \leq 0.05$).

#### 3.3.2. Morphological Observation of Bio-Char by Scanning Electron Microscopy

Figure 4 shows scanning electron microscope (SEM) images of raw biomass and bio-char products obtained at pyrolysis temperatures of 400, 500, 600, 700, and 800 °C. It can clearly be observed that the varying conditions of different pyrolysis temperatures played a significant role in changing the surface morphology of different solid products. The raw biomass appeared to be stone shaped and had a nonporous surface (Figure 4a), and some scattered zones were also observed. The solid bio-char obtained at 400 °C had a rigid, uneven surface with formation of a few pores (Figure 4b). A significant shear bond was witnessed that resulted in the shearing of the materials. The surface of the bio-char product obtained at 500 °C contained a sheet surface and an increased surface area (Figure 4c). The product obtained at 600 °C had no apparent difference in the visual observation (Figure 4d), with only some cracks developing on the surface, when compared to the bio-char that was obtained at 500 °C. The bio-char acquired at 700 °C showed a miscellaneous range of shapes in the pores (Figure 4e). The solid product achieved at 800 °C had a gnarled surface, including dispersed fragments of numerous dimensions (Figure 4f). The advanced carbon content influenced the practice of bio-char. In addition, in the elemental analysis, some variation of the carbon content in bio-char was noted. Therefore, the major features, i.e., morphological, physical, and chemical, were revealed upon the formulation of bio-char. Important parameters such as pyrolysis temperature were found to be dominant, thus controlling the different physical characteristics of bio-char samples, which consequently resulted in each bio-char being significantly different from each other.

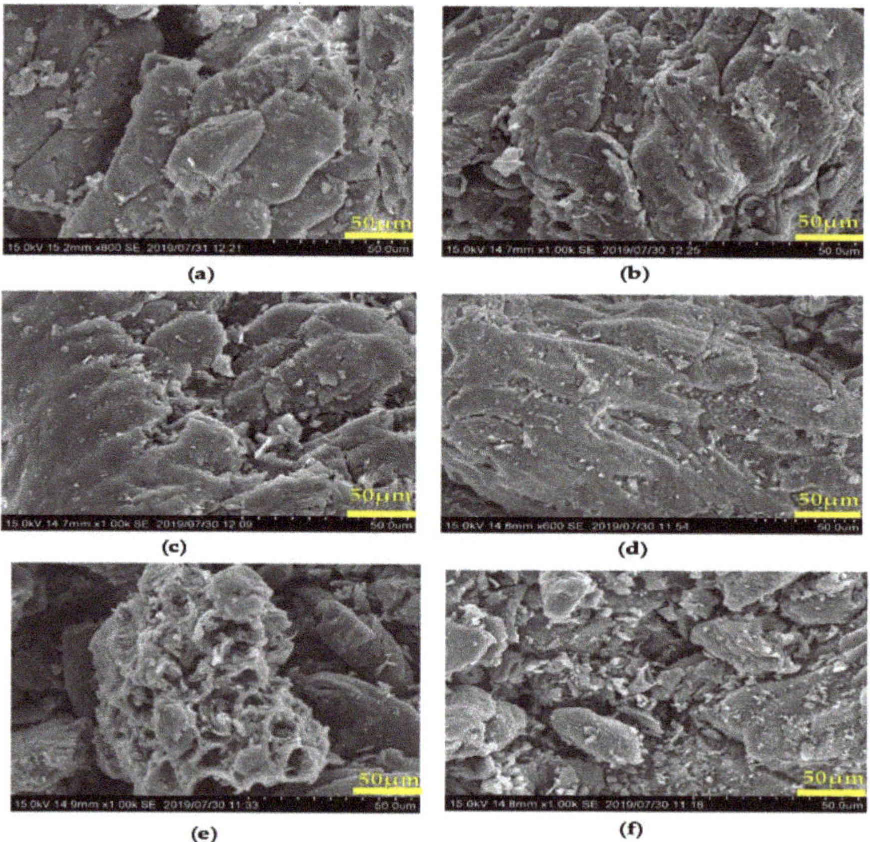

**Figure 4.** SEM results from coconut shells: (**a**) coconut shell and bio-char, (**b**) 400 °C, (**c**) 500 °C, (**d**) 600 °C, (**e**) 700 °C, and (**f**) 800 °C.

### 3.3.3. Higher Heating Value (HHV) of Bio-Char

HHV is an important parameter that defines the efficiency of bio-char as fuel. The HHV did not differ across the temperature ranges (C: $F_{4,10} = 2.09$, $p = 0.156$) (Figure 5), but the HHV at 400 °C was found to be low at 700 and 800 °C ($p \leq 0.05$). However, the HHV showed an increasing trend with the increment of temperature. In this study, the HHV of the bio-chars ranged from 28.1 to 30.6 MJ kg$^{-1}$ for coconut shell when the pyrolysis temperature ranged from 400 to 800 °C. The HHV of the bio-chars increased with the rise in the pyrolysis temperature, as shown in Figure 5. This trend enhanced the carbon content of bio-char at increased temperatures. The HHV of the bio-chars was analogous, in this observation, to several other bio-chars, for example, *Cynara cardunculus L* [35], cotton stalk [36], and cotton stalk briquette [37].

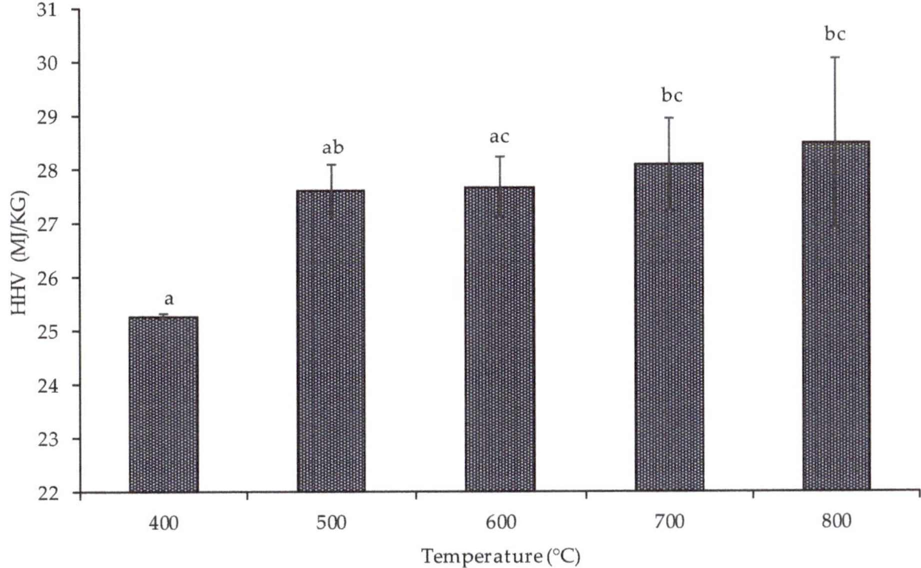

**Figure 5.** Effect of temperature on HHV in coconut shell bio-char, where each bar indicates the mean ± standard error [*SE* (*n* = 3)]. Note: distinct letters in each specific yield bar indicate significant differences (*p* ≤ 0.05).

The HHVs of the bio-chars were equivalent to solid fuels, which are listed from lignite to anthracite, and implies that the bio-chars could be utilized as solid fuels [38].

### 3.4. Characterization of Bio-Oil Using FTIR

Figure 6 shows the FTIR spectra of bio-oil products obtained at 400, 500, 600, 700, and 800 °C pyrolysis temperatures. The spectra were practically indistinguishable and showed that the characteristic formation of liquid product was fetterless in relation to the pyrolysis temperature. However, reliant on the formation of the biomass explanation, the characteristic structure of biomass was described by its components, cellulose, hemicellulose, lignin, for example. These components were found to be identical during the pyrolysis of coconut shells. Moreover, it is common that the bio-oil product predominantly formed from the evaporated volatiles that came from holocellulose (cellulose and hemicellulose), which decomposes at temperatures of approximately 400 °C. Hence, the spectra achieved at several pyrolysis temperatures were analogous, as expected. However, similar outcomes were also observed for the pyrolysis of almond shell [39]. The maximum spacious and comprehensive peak values were observed at 3448 cm$^{-1}$, which was found to be indicative of the excessive existence of oxygenated compounds because it was affected by the O-H extended shock of the hydroxyl groups that exist in water, phenol, alcohol, and/or carboxylic acids. The peak was revealed at 2960 cm$^{-1}$, on account of the C-H outstretched tremor of aliphatic $CH_2$ and $CH_3$ groups [40]. The infirm peak at 2624 cm$^{-1}$ was possibly due to the OH tremor of carboxylic acid. The peak revealed at 2072 cm$^{-1}$ was due to the C=C outstretched terminal alkyne groups. The penetrating peak at 1715 cm$^{-1}$ was due to C=O outstretched tremor, which exposed the availability of aldehydes or ketones. The peak at 1689 cm$^{-1}$ presumably involved the C=C outstretched tremor of alkenes and aromatics. The peaks among 1391 cm$^{-1}$ can be imputed to the C-H bending tremor introduction of alkane groups. The manifest peaks found within the 1300 to 1000 cm$^{-1}$ band were on account of the C=O outstretched and O-H deformation tremors, and showed the availability of alcohols, phenols, esters, and ethers. Furthermore, the peaks lower than 1000 cm$^{-1}$ were distinguishable from polycyclic aromatic compounds and single ring aromatic compounds [41].

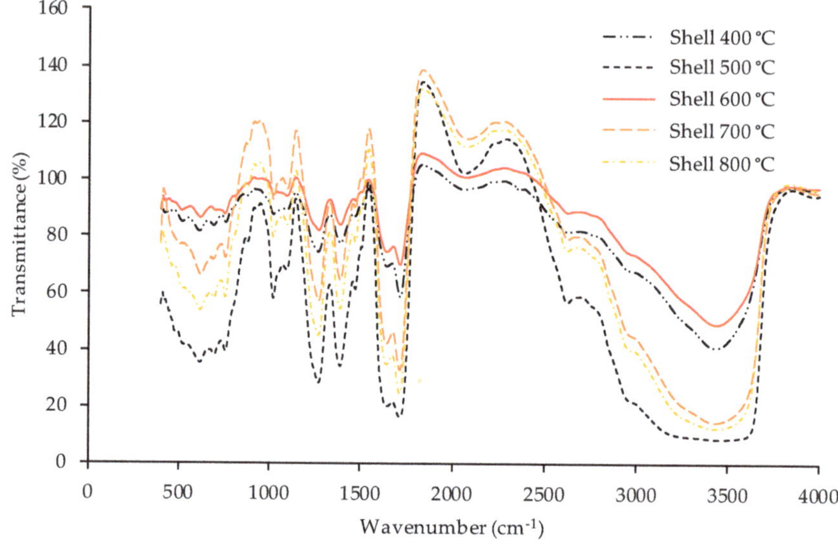


**Figure 6.** Fourier transform infrared spectroscopy (FTIR) spectra of waste coconut shell bio-oil at different temperatures.

### 3.5. Gas Product Characteristics Using Gas Chromatography

Biomass pyrolysis gas was composed of $CO_2$, CO, $CH_4$, $H_2$, and a small amount of hydrocarbons. The gaseous product varied significantly across the temperature range (CO: $F_{4,10} = 158.81$, $p \leq 0.0001$; $CO_2$: $F_{4,10} = 23.74$, $p \leq 0.0001$; $CH_4$: $F_{4,10} = 73.83$, $p \leq 0.0001$; and $H_2$: $F_{4,10} = 492.19$, $p \leq 0.0001$) (Figure 7). The $CO_2$ at 400 and 800 °C significantly differed in all temperature ranges ($p \leq 0.0001$), but no differences were observed at 500, 600, and 700 °C ($p \geq 0.05$). However, CO differed across the temperature ranges ($p \leq 0.0001$) and it increased linearly with increasing temperatures (Figure 7). The formation of non-condensable gas was predominantly the result of secondary reactions, for instance, of the volatile breakdown and the interactions amid volatiles with char or volatiles with gas during pyrolysis; however, CO and $CO_2$ were the prevalent gases. Yang et al. [42] stated that the primary gases of biomass pyrolysis were CO, $CO_2$, $H_2$, $CH_4$, a few organics, and water vapor. Additionally, those gases were significantly responsible in originating the reaction characteristics between the functional groups. For example, $CO_2$ and CO were probably generated below 600 °C by the breakdown and reconstruction of the following functional groups: carboxylic acid (-COOH), carbonyl (C=O), ether (C-O-C), and methane ($CH_4$), which were predominantly generated by the breakdown of O-$CH_3$ groups.

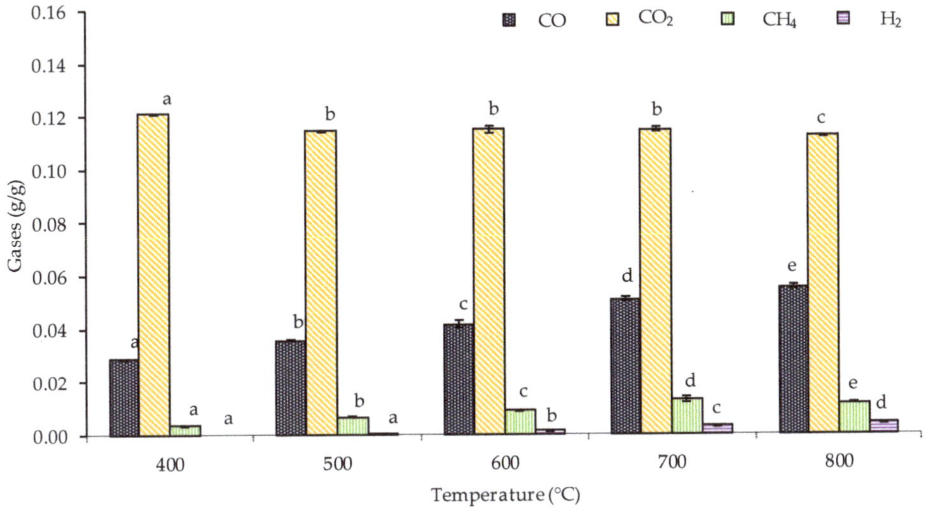


**Figure 7.** Gas yields as a function of pyrolysis temperature using coconut shell. Each bar is the mean ± stand error [*SE* (*n* = 3)]. Different letters in each specific gas bar indicate significant differences ($p \leq$ 0.05).

Figure 7 shows the main components of the gases at different pyrolysis temperatures. When the pyrolysis temperature was 400 °C, then the major components of the gases were CO (31.7 vol.%), $CO_2$ (33.28 vol.%), $H_2$ (2.1 vol.%), and $CH_4$ (7.1 vol.%). By further increasing the pyrolysis temperature, the quantity of $H_2$ quickly rose from 2.1 vol.% (400 °C) to 3.7 vol.% (600 °C), and the $CO_2$ amount decreased from 33.18 vol.% (400 °C) to 26.6 vol.% (600 °C). The higher amounts of the non-condensable gas products were mainly CO and $CO_2$ because of the higher degrees of the deoxygenation component during the biomass pyrolysis experiments. The decrement of $CO_2$ with the increment of temperature in the pyrolysis gas products could be the reason for the gasification reaction of carbon. In addition, by raising the temperature from 600 to 800 °C, the amount of $H_2$ likely increased and the quantity of $CO_2$ likely decreased. The quantity of CO abated gradually with the rise in temperature. With the further increase of temperature, the quantity of $CH_4$ followed a declining trend, indicating that the higher temperature may be significant by promoting dry reforming reactions of CH, along with the decomposition process [43,44].

$$CH_4 + CO_2 = 2CO + 2H_2, \Delta H_{298} = 247.9 \text{ kJ mol}^{-1} \tag{4}$$

$$CH_4 = C + 2H_2, \Delta H_{298} = 75.6 \text{ kJ mol}^{-1} \tag{5}$$

Hydrogen emanated from the dehydrogenation reactions of liquid (bio-oil) and char, for example, alkene formation, condensation, and aromatization. The properties of the gas artifacts varied with temperature; they depended on the development of diverse gases, and were significantly influenced by temperature for their development [45].

Depending on the extensive analyses of the pyrolysis product allocation and development, temperature was significant for fixed bed pyrolysis. High temperature assisted the generation of non-condensable gas properties and the proportion of flammable gas to exhaustive gas production, particularly $H_2$ production [46].

### 3.6. Future Application of Waste Coconut Shell Pyrolysis Product

In recent years, the pyrolysis of biomass has acquired increasing attention. Bio-char is not just suitable for fuel, but can also be additionally prepared into organic fertilizers, electrode materials, or activated carbon [46–48]. Bio-oil is simple to move and use, and is viewed as a perfect energy source that can be applicable in the same way as fossil fuels. Moreover, bio-oil contains a rich mixture of dense organic compounds, which are of great economic value. Upon purification, bio-oil may be utilized to produce valuable chemicals. Non-condensable gas probably underwent combustion in boilers and generators, and likewise, can be utilized for power restoration or to produce syngas. Furthermore, it can be utilized as a supporting agent in the pyrolysis process by supplying heat in the pyrolysis procedure of biomass.

### 4. Conclusions

The pyrolysis reaction temperature had a significant outcome on product characteristics. Developing pyrolysis temperature supports the creation of CO and $CH_4$, and augments the carbon quantity of bio-char; however, it reduced the quantity of water in bio-oil. The ratio of bio-char to the exhaustive quantity of coconut shell samples waned when the pyrolysis temperature rose from 400 to 600 °C. With the continuous raising of the temperature, no momentous differences in the solid bio-char yields were observed. When the pyrolysis temperatures rose from 400 to 600 °C, the product properties of the bio-oil also increased. Likewise, with an increase in the pyrolysis temperature, the amount of bio-oil declined. Pointedly, the weight yield of bio-oil acquired its highest value at 600 °C under all the considered conditions of temperature. Moreover, the existence of higher oxygenated combinations within the liquid product against the temperature ranges considered was confirmed by FTIR analysis. With the increase in the gas volume, there was a more substantial outcome of hydrogen production at higher temperatures, mainly because of the secondary cracking attitudes of the volatiles. Moreover, the temperature had a vital effect on hydrogen ($H_2$) production in waste coconut shell pyrolysis. Bio-char, bio-oil, and non-condensable gaseous products were vital outcomes in the pyrolysis process of biomass, which could be utilized as an alternative energy source like fossil fuels.

**Author Contributions:** Conceptualization, J.K.S. and Q.W.; methodology, Q.W.; software, J.K.S.; validation, J.K.S. and Q.W.; formal analysis, J.K.S.; investigation, J.K.S.; resources, Q.W.; data curation, J.K.S.; writing—original draft preparation, J.K.S.; writing—review and editing, Q.W.; visualization, J.K.S.; supervision, Q.W.; project administration, Q.W.; funding acquisition, Q.W. All authors have read and agreed to the published version of the manuscript.

**Funding:** This study was partially supported by the special funds for Basic Researches (B) (No. 15H05119, FY2015~FY2017) of Grant-in-Aid Scientific Research of the Japanese Ministry of Education, Culture, Sports, Science and Technology (MEXT), Japan.

**Acknowledgments:** The authors would like to thank Abedur Rahman (Doctoral candidate, Hydraulic and Environmental Engineering Laboratory, Department of Civil and Environmental Engineering, Saitama University, Saitama, Japan) and Nazim Uddin (PhD Research Fellow Ecology Group, Department of Environmental Science Victoria University, Melbourne, Australia) for providing technical support and discussions. Moreover, first author would like to thank Mitsubishi Corporation for providing a scholarship that made it possible to achieve this study.

**Conflicts of Interest:** The authors declare no conflict of interest.

### References

1. Enweremadu, C.C.; Mbarawa, M.M. Technical aspects of production and analysis of biodiesel from used cooking oil—A review. *Renew. Sustain. Energy Rev.* **2009**, *13*, 2205–2224. [CrossRef]
2. Biswas, S.; Majhi, S.; Mohanty, P.; Pant, K.K.; Sharma, D.K. Effect of different catalyst on the co-cracking of Jatropha oil, vacuum residue and high density polyethylene. *Fuel* **2014**, *133*, 96–105. [CrossRef]
3. Goyal, H.B.; Seal, D.; Saxena, R.C. Bio-fuels from thermochemical conversion of renewable resources: A review. *Renew. Sustain. Energy Rev.* **2008**, *12*, 504–517. [CrossRef]
4. Khan, A.A.; De Jong, W.; Jansens, P.J.; Spliethoff, H. Biomass combustion in fluidized bed boilers: Potential problems and remedies. *Fuel Process. Technol.* **2009**, *90*, 21–50. [CrossRef]

5. Park, H.J.; Dong, J.I.; Jeon, J.K.; Park, Y.K.; Yoo, K.S.; Kim, S.S.; Kim, J.; Kim, S. Effects of the operating parameters on the production of bio-oil in the fast pyrolysis of Japanese larch. *Chem. Eng. J.* **2008**, *143*, 124–132. [CrossRef]

6. Mašek, O.; Budarin, V.; Gronnow, M.; Crombie, K.; Brownsort, P.; Fitzpatrick, E.; Hurst, P. Microwave and slow pyrolysis biochar—Comparison of physical and functional properties. *J. Anal. Appl. Pyrolysis* **2013**, *100*, 41–48. [CrossRef]

7. Fu, P.; Hu, S.; Xiang, J.; Sun, L.; Su, S.; Wang, J. Evaluation of the porous structure development of chars from pyrolysis of rice straw: Effects of pyrolysis temperature and heating rate. *J. Anal. Appl. Pyrolysis* **2012**, *98*, 177–183. [CrossRef]

8. Mohanty, P.; Pant, K.K.; Naik, S.N.; Parikh, J.; Hornung, A.; Sahu, J.N. Synthesis of green fuels from biogenic waste through thermochemical route–The role of heterogeneous catalyst: A review. *Renew. Sustain. Energy Rev.* **2014**, *38*, 131–153. [CrossRef]

9. McHenry, M.P. Agricultural bio-char production, renewable energy generation and farm carbon sequestration in Western Australia: Certainty, uncertainty and risk. *Agric. Ecosyst. Environ.* **2009**, *129*, 1–7. [CrossRef]

10. Malghani, S.; Gleixner, G.; Trumbore, S.E. Chars produced by slow pyrolysis and hydrothermal carbonization vary in carbon sequestration potential and greenhouse gases emissions. *Soil Biol. Biochem.* **2013**, *62*, 137–146. [CrossRef]

11. Hao, X.W.; Huang, Y.Z.; Cui, Y.S. Effect of bone char addition on the fractionation and bio-accessibility of Pb and Zn in combined contaminated soil. *Acta Ecol. Sin.* **2010**, *30*, 118–122. [CrossRef]

12. Khare, P.; Goyal, D.K. Effect of high and low rank char on soil quality and carbon sequestration. *Ecol. Eng.* **2013**, *52*, 161–166. [CrossRef]

13. Rout, T.; Pradhan, D.; Singh, R.K.; Kumari, N. Exhaustive study of products obtained from coconut shell pyrolysis. *J. Environ. Chem. Eng.* **2016**, *4*, 3696–3705. [CrossRef]

14. Kim, S.W. Pyrolysis conditions of biomass in fluidized beds for production of bio-oil compatible with petroleum refinery. *J. Anal. Appl. Pyrolysis* **2016**, *117*, 220–227. [CrossRef]

15. Le Brech, Y.; Jia, L.; Cissé, S.; Mauviel, G.; Brosse, N.; Dufour, A. Mechanisms of biomass pyrolysis studied by combining a fixed bed reactor with advanced gas analysis. *J. Anal. Appl. Pyrolysis* **2016**, *117*, 334–346. [CrossRef]

16. Santos, R.M.; Santos, A.O.; Sussuchi, E.M.; Nascimento, J.S.; Lima, Á.S.; Freitas, L.S. Pyrolysis of mangaba seed: Production and characterization of bio-oil. *Bioresour. Technol.* **2015**, *196*, 43–48. [CrossRef]

17. Gu, X.; Ma, X.; Li, L.; Liu, C.; Cheng, K.; Li, Z. Pyrolysis of poplar wood sawdust by TG-FTIR and Py–GC/MS. *J. Anal. Appl. Pyrolysis* **2013**, *102*, 16–23. [CrossRef]

18. Raveendran, K.; Ganesh, A.; Khilar, K.C. Influence of mineral matter on biomass pyrolysis characteristics. *Fuel* **1995**, *74*, 1812–1822. [CrossRef]

19. Hoque, M.M.; Bhattacharya, S.C. Fuel characteristics of gasified coconut shell in a fluidized and a spouted bed reactor. *Energy* **2001**, *26*, 101–110. [CrossRef]

20. Siengchum, T.; Isenberg, M.; Chuang, S.S. Fast pyrolysis of coconut biomass–An FTIR study. *Fuel* **2013**, *105*, 559–565. [CrossRef]

21. Tsamba, A.J.; Yang, W.; Blasiak, W. Pyrolysis characteristics and global kinetics of coconut and cashew nut shells. *Fuel Process. Technol.* **2006**, *87*, 523–530. [CrossRef]

22. Raj, K.G.; Joy, P.A. Coconut shell based activated carbon–iron oxide magnetic nanocomposite for fast and efficient removal of oil spills. *J. Environ. Chem. Eng.* **2015**, *3*, 2068–2075. [CrossRef]

23. Tsai, W.T.; Lee, M.K.; Chang, Y.M. Fast pyrolysis of rice straw, sugarcane bagasse and coconut shell in an induction-heating reactor. *J. Anal. Appl. Pyrolysis* **2006**, *76*, 230–237. [CrossRef]

24. Sundaram, E.G.; Natarajan, E. Pyrolysis of coconut shell: An experimental investigation. *J. Eng. Res.* **2009**, *6*, 33–39. [CrossRef]

25. Sait, H.H.; Hussain, A.; Salema, A.A.; Ani, F.N. Pyrolysis and combustion kinetics of date palm biomass using thermogravimetric analysis. *Bioresour. Technol.* **2012**, *118*, 382–389. [CrossRef] [PubMed]

26. Sarenbo, S. Wood ash dilemma-reduced quality due to poor combustion performance. *Biomass Bioenergy* **2009**, *33*, 1212–1220. [CrossRef]

27. Kan, T.; Strezov, V.; Evans, T.J. Lignocellulosic biomass pyrolysis: A review of product properties and effects of pyrolysis parameters. *Renew. Sustain. Energy Rev.* **2016**, *57*, 1126–1140. [CrossRef]

28. Damartzis, T.; Vamvuka, D.; Sfakiotakis, S.; Zabaniotou, A. Thermal degradation studies and kinetic modeling of cardoon (Cynara cardunculus) pyrolysis using thermogravimetric analysis (TGA). *Bioresour. Technol.* **2011**, *102*, 6230–6238. [CrossRef]

29. Zhao, X.; Wang, M.; Liu, H.; Li, L.; Ma, C.; Song, Z. A microwave reactor for characterization of pyrolyzed biomass. *Bioresour. Technol.* **2012**, *104*, 673–678. [CrossRef]

30. Chen, D.; Liu, D.; Zhang, H.; Chen, Y.; Li, Q. Bamboo pyrolysis using TG–FTIR and a lab-scale reactor: Analysis of pyrolysis behavior, product properties, and carbon and energy yields. *Fuel* **2015**, *148*, 79–86. [CrossRef]

31. Kim, K.H.; Eom, I.Y.; Lee, S.M.; Choi, D.; Yeo, H.; Choi, I.G.; Choi, J.W. Investigation of physicochemical properties of biooils produced from yellow poplar wood (Liriodendron tulipifera) at various temperatures and residence times. *J. Anal. Appl. Pyrolysis* **2011**, *92*, 2–9. [CrossRef]

32. Yang, H.; Yan, R.; Chen, H.; Lee, D.H.; Liang, D.T.; Zheng, C. Mechanism of palm oil waste pyrolysis in a packed bed. *Energy Fuels* **2006**, *20*, 1321–1328. [CrossRef]

33. Demirbaş, A. Mechanisms of liquefaction and pyrolysis reactions of biomass. *Energy Convers. Manag.* **2000**, *41*, 633–646. [CrossRef]

34. Becidan, M.; Skreiberg, Ø.; Hustad, J.E. NOx and N2O Precursors (NH3 and HCN) in Pyrolysis of Biomass Residues. *Energy Fuels* **2007**, *21*, 1173–1180. [CrossRef]

35. Encinar, J.M.; Gonzalez, J.F.; Gonzalez, J. Fixed-bed pyrolysis of Cynara cardunculus L. Product yields and compositions. *Fuel Process. Technol.* **2000**, *68*, 209–222. [CrossRef]

36. Chen, Y.; Yang, H.; Wang, X.; Zhang, S.; Chen, H. Biomass-based pyrolytic polygeneration system on cotton stalk pyrolysis: Influence of temperature. *Bioresour. Technol.* **2012**, *107*, 411–418. [CrossRef]

37. Yang, H.; Liu, B.; Chen, Y.; Chen, W.; Yang, Q.; Chen, H. Application of biomass pyrolytic polygeneration technology using retort reactors. *Bioresour. Technol.* **2016**, *200*, 64–71. [CrossRef]

38. Raveendran, K.; Ganesh, A. Heating value of biomass and biomass pyrolysis products. *Fuel* **1996**, *75*, 1715–1720. [CrossRef]

39. Gonzalez, J.F.; Ramiro, A.; González-García, C.M.; Gañán, J.; Encinar, J.M.; Sabio, E.; Rubiales, J. Pyrolysis of almond shells. Energy applications of fractions. *Ind. Eng. Chem. Res.* **2005**, *44*, 3003–3012. [CrossRef]

40. Apaydin-Varol, E.; Uzun, B.B.; Önal, E.; Pütün, A.E. Synthetic fuel production from cottonseed: Fast pyrolysis and a TGA/FT-IR/MS study. *J. Anal. Appl. pyrolysis* **2014**, *105*, 83–90. [CrossRef]

41. Demiral, I.; Şensöz, S. The effects of different catalysts on the pyrolysis of industrial wastes (olive and hazelnut bagasse). *Bioresour. Technol.* **2008**, *99*, 8002–8007. [CrossRef] [PubMed]

42. Yang, H.; Yan, R.; Chen, H.; Lee, D.H.; Zheng, C. Characteristics of hemicellulose, cellulose and lignin pyrolysis. *Fuel* **2007**, *86*, 1781–1788. [CrossRef]

43. Domínguez, A.; Fernández, Y.; Fidalgo, B.; Pis, J.J.; Menéndez, J.A. Biogas to syngas by microwave-assisted dry reforming in the presence of char. *Energy Fuels* **2007**, *21*, 2066–2071. [CrossRef]

44. Dominguez, A.; Menéndez, J.A.; Fernandez, Y.; Pis, J.J.; Nabais, J.V.; Carrott, P.J.M.; Carrott, M.R. Conventional and microwave induced pyrolysis of coffee hulls for the production of a hydrogen rich fuel gas. *J. Anal. Appl. Pyrolysis* **2007**, *79*, 128–135. [CrossRef]

45. Yang, H.; Yan, R.; Chen, H.; Lee, D.H.; Liang, D.T.; Zheng, C. Pyrolysis of palm oil wastes for enhanced production of hydrogen rich gases. *Fuel Process. Technol.* **2006**, *87*, 935–942. [CrossRef]

46. Dufour, A.; Girods, P.; Masson, E.; Rogaume, Y.; Zoulalian, A. Synthesis gas production by biomass pyrolysis: Effect of reactor temperature on product distribution. *Int. J. Hydrog. Energy* **2009**, *34*, 1726–1734. [CrossRef]

47. Chen, D.; Zhou, J.; Zhang, Q. Effects of heating rate on slow pyrolysis behavior, kinetic parameters and products properties of moso bamboo. *Bioresour. Technol.* **2014**, *169*, 313–319. [CrossRef]

48. Lee, Y.; Park, J.; Ryu, C.; Gang, K.S.; Yang, W.; Park, Y.K.; Jung, J.; Hyun, S. Comparison of biochar properties from biomass residues produced by slow pyrolysis at 500 °C. *Bioresour. Technol.* **2013**, *148*, 196–201. [CrossRef]

*Article*

# Lipid Production from Amino Acid Wastes by the Oleaginous Yeast *Rhodosporidium toruloides*

Qiang Li [1,2,†], Rasool Kamal [1,2,†], Qian Wang [1,3], Xue Yu [1,3] and Zongbao Kent Zhao [1,3,*]

[1]   Division of Biotechnology, Dalian Institute of Chemical Physics, Chinese Academy of Sciences, Dalian 116023, China; leeq@dicp.ac.cn (Q.L.); kamal@dicp.ac.cn (R.K.); wangqian@dicp.ac.cn (Q.W.); dlyx0213@dicp.ac.cn (X.Y.)
[2]   School of Chemical Engineering, University of Chinese Academy of Sciences, Beijing 100049, China
[3]   Dalian Key Laboratory of Energy Biotechnology, Dalian Institute of Chemical Physics, Chinese Academy of Sciences, Dalian 116023, China
*   Correspondence: zhaozb@dicp.ac.cn; Tel.: +86-0411-8437-9211
†   These authors contributed equally.

Received: 25 February 2020; Accepted: 21 March 2020; Published: 1 April 2020

**Abstract:** Microbial lipids have been considered as promising resources for the production of renewable biofuels and oleochemicals. Various feedstocks, including sugars, crude glycerol, and volatile fatty acids, have been used as substrates for microbial lipid production, yet amino acid (AA) wastes remain to be evaluated. Here, we describe the potential to use AA wastes for lipid production with a two-stage culture mode by an oleaginous yeast strain *Rhodosporidium toruloides* CGMCC 2.1389. Each of the 20 proteinogenic AAs was evaluated individually as sole carbon source, with 8 showing capability to facilitate cellular lipid contents of more than 20%. It was found that L-proline was the most favored AA, with which cells accumulated lipids to a cellular lipid content of 37.3%. When blends with AA profiles corresponding to those of meat industry by-products and sheep viscera were used, the cellular lipid contents reached 27.0% and 28.7%, respectively. The fatty acid compositional analysis of these lipid products revealed similar profiles to those of vegetable oils. These results, thus, demonstrate a potential route to convert AA wastes into lipids, which is of great importance for waste management and biofuel production.

**Keywords:** amino acid wastes; biofuels; microbial lipids; *Rhodosporidium toruloides*; two-stage culture

---

## 1. Introduction

Biodiesel has emerged as one of the most promising energy sources for the renewable biofuel market, owing to its excellent compatibility with the current fuel infrastructure systems [1]. Lipids produced by oleaginous microorganisms have been exploited as alternative feedstocks for biodiesel production [2]. Some oleaginous yeasts are attractive due to their high growth rates and high cellular lipid contents [3]. Specifically, the oleaginous yeast *Rhodosporidium toruloides*, recently reclassified as *Rhodotorola toruloides*, has been demonstrated as an excellent lipid producer because it can accumulate large amounts of lipids under high cell-density culture conditions [4–7]. More importantly, *R. toruloides* uses diverse substrates for lipid production and can naturally tolerate inhibitory compounds found in hydrolysates of lignocellulosic biomass [8]. Although various raw materials have been utilized for microbial lipid production, including carbohydrates from various sources, waste glycerol, and volatile fatty acids [9], the costs of feedstocks remain remarkably high to ensure economic competitiveness of microbial lipids to vegetable oils [10]. Therefore, efforts are devoted continuously to exploring innovative processes, new feedstocks, and valuable co-products in order to improve the techno-economics of microbial lipid technology.

Amino acid (AA) wastes have been implicated as possible feedstocks for the production of biochemicals and biofuels, such as biobutanol from engineered microorganisms [11,12]. In fact, AAs can be envisioned as organic amines; once the nitrogen atom is released from the AA, the residual carbon skeleton can be readily converted into pyruvate or an intermediate of the citric acid cycle, thus fueling cellular metabolism [13]. Previously, food waste hydrolysates were evaluated for lipid and protein production by *R. toruloides* Y2 [14]. So far, AA wastes have not been used alone for microbial lipid production, likely because oleaginous microorganisms normally accumulate lipids under nitrogen limitation [15], while the catabolism of AA naturally generates a relatively nitrogen-rich environment. Interestingly, early studies indicated that either phosphate limitation or a two-stage culture mode could be used to achieve lipid production under nitrogen-rich conditions [16,17] or with nitrogen-containing substrates, such as chitin degradation products [18]. The meat wastes generated from abattoirs and meat processing industries correspond up to 50% of the total slaughtered animal weight, which is costly in terms of ecological disposal [19]. Specifically, huge amounts of fish and sheep wastes are produced [20,21], and these protein wastes can be further converted into AA mixtures [22]. Unlawful disposal of these wastes is known to cause serious environmental problems [23]. Therefore, the conversion of AA wastes into lipids merits some efforts in terms of investigating meat industry profitability, wastes management, and biofuel production.

The aim of this study was to explore the potential to use AA wastes for lipid production by the oleaginous yeast *Rhodosporidium toruloides* CGMCC 2.1389. The carbon sources of the lipid production cultures were made with each proteinogenic AA alone, or with designated blends with AA profiles corresponding to those of fish muscle (FM) [24], meat industry by-products (MI) [25], or sheep viscera (SV) [21], as these are the major meat wastes with abundant proteins and AA contents [20]. Results showed that *R. toruloides* CGMCC 2.1389 could accumulate lipids to more than 20% when cultivated with some AA alone or blends as sole carbon sources by using a two-stage culture mode [26]. Further analysis indicated that those neutral lipid products comprised mainly long chain fatty acids with 16 or 18 carbon atoms, which may be used to make biodiesel and other related oleochemicals. This study demonstrates that AA and related nitrogen-rich wastes can be explored to produce microbial lipids, which fits well with the protein-based biorefinery concept [12].

## 2. Materials and Methods

### 2.1. Microorganism, Media, and Growth Conditions

The yeast strain *R. toruloides* CGMCC 2.1389, originally obtained from China General Microbiology Collection Center, was maintained at 4 °C on yeast extract-peptone-dextrose (YEPD) agar plate contained (g/L) glucose·$H_2O$ 20, peptone 10, yeast extract 10, and agar 20, and was sub-cultured twice a month. The peptone (total nitrogen 14.5%) and yeast extract (total nitrogen 9.0%) were obtained from Aoboxing Biotech. Co. Ltd. (Beijing, China).

The medium used for seed culture contained (g/L) glucose·$H_2O$ 20, yeast extract 10, and peptone 10 (pH 6.0). For lipid production experiments, media with single AA or AA blends in 500 mM 2-(N-morpholino) ethanesulfonate (MES) buffer (pH 5.5) were used, and the concentrations of AAs were adjusted such that the media contained a total carbon at 16 or 28 g/L (unless otherwise specified). Accordingly, media with a single AA contained (g/L) L-asparagine (Asn) 50, L-aspartic acid (Asp) 44.66, L-valine (Val) 31.23, L-isoleucine (Ile) 29.15, L-arginine (Arg) 46.81, L-methionine (Met) 39.79, L-glutamine (Gln) 38.97, L-histindine (His) 42.58, L-glutamic acid (Glu) 39.2, L-proline (Pro) 30.6, L-alanine (Ala) 69.26, L-serine (Ser) 46.7, L-threonine (Thr) 39.7, L-glycine (Gly) 50, L-phenylalanine (Phe) 24.46, L-cysteine (Cys) 78, L-tryptophan (Trp) 24.75, L-lysine (Lys) 40.59, L-tyosine (Tyr) 26.8, or L-leucine (Leu) 29.15. Media with AA blends contained mixtures, with their compositions shown in Table 1. All the media were sterilized at 121 °C for 20 min.

**Table 1.** Compositional profiles of amino acid (AA) blends used for lipid production.

| AA | Initial Concentration (g/L) | | |
|---|---|---|---|
| | SV Blends | FM Blends | MI Blends |
| L-Aspartic acid (Asp) | 3.22 | 1.14 | 5.02 |
| DL-Asparagine (Asn) | 3.21 | - | - |
| L-Isoleucine (Ile) | 2.62 | 2.54 | 2.32 |
| L-Valine (Val) | 3.45 | 3.07 | 2.90 |
| L-Methionine (Met) | 0.82 | 2.51 | 1.36 |
| L-Arginine (Arg) | - | 3.47 | 4.15 |
| L-Histidine (His) | 1.46 | 1.16 | 1.64 |
| L-Glutamine (Gln) | 3.82 | - | - |
| L-Proline (Pro) | 4.56 | 0.43 | 8.77 |
| L-Glutamic acid (Glu) | 7.30 | 6.50 | 8.96 |
| L-Alanine (Ala) | 5.39 | 3.12 | 3.96 |
| L-Threonine (Thr) | 2.69 | 6.54 | 2.51 |
| L-Glycine (Gly) | 7.95 | 1.20 | 6.17 |
| L-Serine (Ser) | 3.18 | 2.01 | 2.90 |
| L-Cysteine (Cys) | 0.82 | 1.49 | 0.66 |
| L-Phenylalanine (Phe) | 3.65 | 6.46 | 2.22 |
| L-Tryptophan (Trp) | - | 2.13 | 0.65 |
| L-Tyrosine (Tyr) | 0.73 | 5.91 | 1.64 |
| L-Lysine (Lys) | 5.06 | 5.85 | 4.73 |
| L-Leucine (Leu) | 5.28 | 4.37 | 3.95 |
| Total | 65.23 | 59.88 | 64.52 |

All amino acids were of analytical grade from Sangon Biotech (Beijing, China), with analytical grade reagents and chemicals purchased locally.

## 2.2. Culture Conditions

*R. toruloides* CGMCC 2.1389 cells were cultivated in YEPD media at 30 °C, 200 rpm for 24 h, then harvested by centrifugation at 5000 rpm for 5 min and washed twice with distilled water. To produce lipids, cells were resuspended in AA media in 500 mL shake flasks at an initial cell density of 4.0 g/L, and incubated at 30 °C and 200 rpm for 108 h, unless otherwise specified.

All culture experiments for lipid production were done in triplicate, and error bars shown in figures are standard deviations.

## 2.3. Analytical Methods

To determine dry cell mass, cells in 30 mL of culture broth were centrifuged at 8000 rpm for 5 min, washed twice with distilled water, dried at 105 °C for 12 h to constant weight, and determined gravimetrically [27]. Lipid was extracted with methanol/chloroform (1:2, v/v) according to a known method [5]. Cellular lipid contents were obtained by dividing lipid with dry cell weight.

Lipid products were transmethylated and analyzed by using a gas chromatography (GC) method [5]. Briefly, lipid samples (70 mg) were stirred with 5% KOH methanol solution (0.5 mL) at 65 °C for 50 min, then 0.7 mL of $BF_3$ diethyl etherate and methanol solution (4:6) were added, refluxed for 10 min, cooled, diluted with distilled water, and extracted with n-hexane. The organic layer was washed twice with distilled water and used for analysis. Finally, the compositional profiling of fatty acids was measured by using a 7890F GC system (Techcomp Scientific Instrument Co. Ltd., Shanghai, China), equipped with a cross-linked capillary free fatty acid phase (FFAP) column (30 mm × 0.25 mm × 0.25 mm) and a flame ionization detector. The flow rates for $N_2$, $H_2$, and air were 720 mL/min, 30 mL/min, and 100 mL/min, respectively. The temperatures of the injection port, oven, and detector were set at 250, 190, and 280 °C, respectively. The injection volume was 0.5 uL. Fatty acids were identified by comparing them with the retention time of standards and quantifying them by the respective peak areas.

AAs were analyzed at 30 °C using a Dionex ICS-5000 ion chromatography system (Thermo-Fisher Scientific, MA, USA). The AminoPac PA10 column set consisting of a guard column (4 mm × 50 mm) and an analytical column (4 mm × 250 mm) was used to separate individual AAs. Gradient elution was performed at a flow-rate of 0.25 mL/min, with water, sodium hydroxide, and sodium acetate as mobile phases. The gradient conditions and the standard chromatogram used to analyze 20 proteinogenic AAs are shown in Supplementary Materials (Table S1 and Figure S1). AAs were quantified based on standard curves obtained under the same chromatographic conditions. Individual standard curves were established based on the correspondence between the AA concentration and peak area.

*2.4. Statistical Analysis*

SPSS Statistics 23 (IBM Software, Inc., California, USA) was used for statistical analysis. Two-way ANOVA with Tukey's multiple comparison test was conducted to compare different groups. Degrees of freedom, sum of squares, mean square, and distribution of the ratio among p-0.05 were taken into consideration, the results of which are shown in Tables S2–S5. Data with $p < 0.05$ was considered statistically significant.

## 3. Results and Discussion

*3.1. Evaluating Individual AAs as Carbon Sources for Lipid Production*

It is well known that the carbon skeletons of AAs upon transamination or deamination can be further converted into metabolites, such as pyruvate, acetyl-CoA, acetoacetyl-CoA, or citric acid cycle intermediates [13,28,29], and those intermediates can be used to synthesize fatty acids and lipids. Two-stage culture mode with an initial cell density of 4.0 g/L was used to evaluate the capability of *R. toruloides* cells to produce lipids on each proteinogenic AA. To make a reasonable comparison, initial AA concentration was set at a total carbon concentration of 16 g/L. It was found that there were cell mass increments for 11 AAs, and most of these cases also had cellular lipid contents higher than 20% (Figure 1). Specifically, lipid contents were 27.3%, 26.3%, 25.3%, 23.5%, 21.7%, and 22.0% for the cultures with Glu, Pro, Ala, Asp, Ser, and Gln, respectively. There were small cell mass changes for Gly and Tyr, and significantly reduced cell mass was observed for Met, His, Arg, Thr, Trp, Tyr, Lys, and Leu ($p < 0.05$).

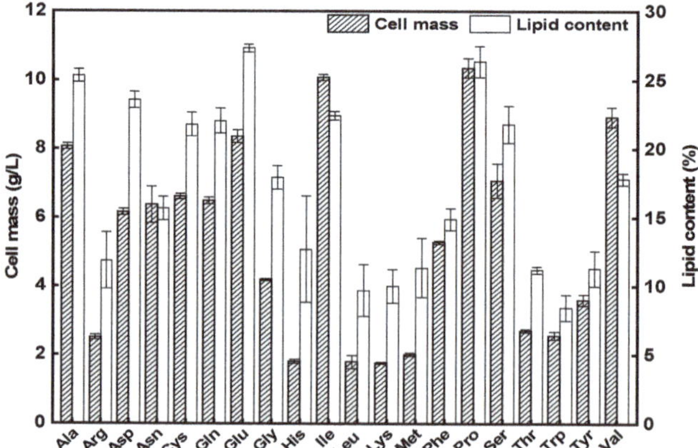


**Figure 1.** Lipid production results for *R. toruloides* CGMCC 2.1389 with each AA.

Literally, AAs can be classified via their catabolic precursors into ketogenic AAs, glucogenic AAs, or both. Similarly, they can be categorized on the basis of their structures and the chemical characteristics of their side chains into aliphatic, hydroxyl, aromatic, acidic, basic, or neutral AAs [28]. So far, substantial differences have been found, even within the same group, in terms of their efficacy as substrates for lipid production by *R. toruloides*. For instance, while aliphatic AAs such as Ala, Ile and Val were favored for lipid production, Leu was disfavored. For hydroxyl-group-containing AAs, Ser was favored, while Thr was disfavored. These results further suggest complex regulation mechanisms for lipid accumulation beyond the carbon sources of catabolism. While detailed discussions about the metabolism of each AA are beyond the scope of this work, these data are valuable references for further engineering of the yeast *R. toruloides* with genetic tools [30].

### 3.2. Lipid Production on AA Blends

Next, we used AA blends as carbon sources for lipid production by *R. toruloides*. Three AA blends were used, which had AA compositions similar to FM, MI, and SV, respectively (Table 1). The amounts of each AA were added such that the total carbon concentration was 28 g/L. Results showed that cells accumulated lipids close to or more than 20% after 108 h (Figure 2a). The lipid contents were 28% and 27%, and cell mass contents were 8.37 g/L and 6.50 g/L, in amino acids form sheep viscera (SVAA) and amino acids form meat industry by-products (MIAA), respectively. In terms of the culture with amino acids form fish muscle (FMAA), the lipid content and cell mass were 19% and 4.45 g/L, respectively. The compositional profiles of these AA blends may have major contributions to the lipid production results. For FMAA media, relatively high contents of Met, His, Arg, Thr, Cys, Trp, Tyr, Lys, and Leu were found, which failed to support lipid accumulation by *R. toruloides* cells when used as sole carbon sources in the media (vide ante). On the other hand, SVAA and MI media contain high amounts of AAs that support lipid accumulation, such as Asp, Pro, Ala, Glu, and Ser, yet low amounts of those disfavoring AAs, such as Met, Thr, Trp, and Tyr (Figure 1).

The initial C/N molar ratios of the media were determined as 4.43, 3.75, and 3.68 for FM, MI, and SV, respectively, based on their AA compositions. It should be noted that such low C/N molar ratios were inadequate to stimulate lipid accumulation, as C/N molar ratios of more than 70 were suggested for high lipid production by oleaginous yeasts [31]. As no phosphates were included in the media, cells were subjected to strict phosphate limitation. Thus, the fact that *R. toruloides* cells in AA media blends accumulated lipids close to or more than 20% was in agreement with the mechanism of phosphate-limitation-induced lipid production [16,17].

We also traced the residual AA at the end of the culture and the results are shown in Figure 2b–d. It was found that some AAs, such as Asp, Asn, Glu, Pro, Glu, Ala, and Ser, were largely consumed when initially included in the media, while others such as Ile, Val, His, Met, Thr, Gly, Cys, Trp, Tyr, Lys, and Leu were less utilized. Met, His, Trp, and Tyr were among the most disfavored ones, as there was little difference between their corresponding initial and residual data. Interestingly, Ala was found, with less than 50% being utilized in the MI media (Figure 2d), but was exhausted in the other two media (Figure 2b,c). Also, Asp was found, with about 30% leftover in the SV media (Figure 2c), but was exhausted in the other two media (Figure 2b,d). The differences in AA utilization patterns showed that *R. toruloides* cells favor some AAs, such as Pro, Glu, Asp, Ala, Asn, and Ser, while disfavoring some others. It should be noted that *R. toruloides* is a wild-type strain with no AA auxotrophic phenotype, indicating a full competence of AA metabolism. Thus, it is most likely that this yeast lacks an effective importing system for those disfavored AAs. However, the physiochemical properties of AAs in the media in a slightly acidic environment may also play a role in their uptake.

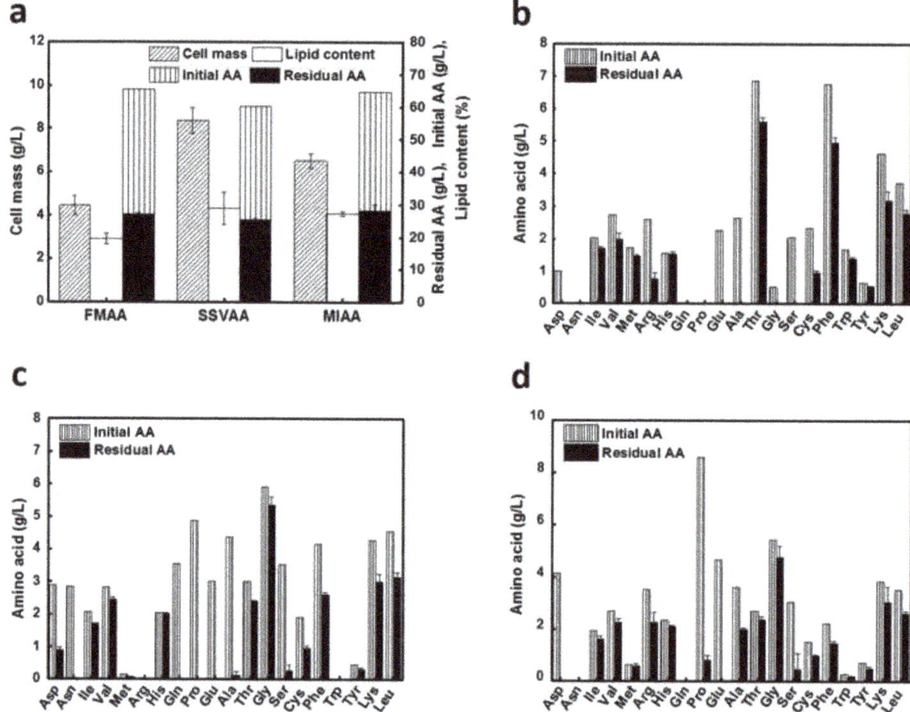

**Figure 2.** Results of lipid production at 30 °C and 200 rpm for 108 h from AA blends at a total carbon concentration of 28 g/L. (**a**) Lipid production in FMAA, SVAA, and MIAA blends. (**b**) Initial and residual AA profiles of FMAA media blend. (**c**) Initial and residual AA profiles of SVAA media blend. (**d**) Initial and residual AA profiles of MIAA media blend.

### 3.3. Lipid Production on L-Proline

The above results showed that L-proline was favored by *R. toruloides*, whether used alone or presented in AA blends. Thus, more experiments were designed to assess its capacity as a carbon source for lipid production by using the two-stage culture approach. Here, initial Pro concentrations of 30.6, 53.6, and 76.6 g/L, corresponding to total carbon concentrations of 16 (Pro-16), 28 (Pro-28), and 40 g/L (Pro-40), respectively, were used in the media. Results showed that there were no major differences among these three groups in terms of cellular lipid content (Figure 3). While cell mass was slightly lower (7.0 g/L) for the Pro-40 group, it was essentially identical for the other two groups. Lipid titers were 2.5, 2.6, and 2.1 g/L for Pro-16, Pro-28, and Pro-40, respectively. Thus, there were no significant differences in terms of lipid production, yet proline consumption was increased significantly with an increase in proline concentration ($p < 0.05$). The proline consumption was 19.8, 29.0, and 43.6 g/L for Pro-16, Pro-28, and Pro-40 media, respectively. These data also indicated that Pro at high initial concentrations may exert more osmotic stress [5,27,32], leading to inhibitory effects. For the Pro-28 media, when the culture time increased to 180 h, the cell mass and lipid content were 8.7 g/L and 37.3%, respectively, and there was 22.4 g/L Pro leftover. It seemed that there were limited benefits with prolonged culture time. Nonetheless, the data confirmed that Pro was a relatively good substrate for lipid production for *R. toruloides*.

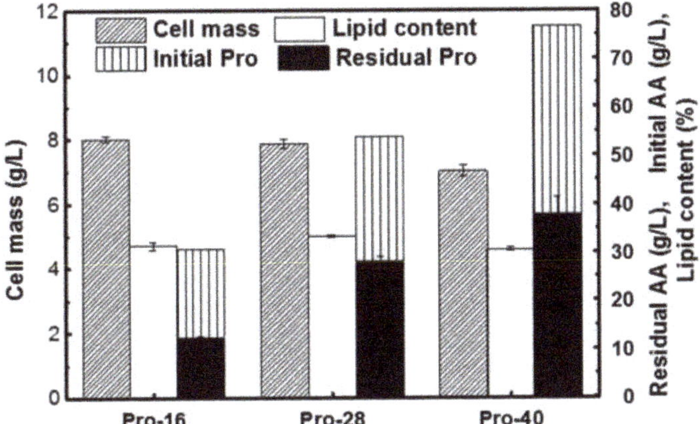



**Figure 3.** Results of lipid production on Pro at different initial concentrations. Cultures were performed at 30 °C and 200 rpm for 108 h.

### 3.4. Fatty Acid Compositional Profile of the Lipid Products

The lipid samples produced on different substrates were transmethylated into fatty acid methyl esters and analyzed by GC. It was found that palmitic acid (16:0), stearic acid (18:1), and oleic acid (18:0) were the major ones, and that no major fatty acid distributional differences were noticed among those products (Table 2). It should be noted that *R. toruloides* could produce lipids with different fatty acid compositional profiles [5,26,33,34]. As noticed in Table 2, lipid products from AA are more similar to lipids produced from corn stover [34] and palm [35]. Nonetheless, microbial lipids carrying long chain fatty acids with 16 and 18 carbons as the major fractions, which are similar to those of conventional vegetable oils form palm and canola, have been considered as alternative feedstock for biodiesel production [36]. Thus, lipids produced on AA have the potential for biodiesel production.

**Table 2.** Fatty acid compositional profiles of lipids produced on different substrates by *R. toruloides* and typical vegetable oils.

| Media | Lipid Content (%) | Relative Fatty Acid Content (%, w/w) | | | | | |
|---|---|---|---|---|---|---|---|
| | | Myristic (14:0) | Palmitic (16:0) | Palmitoleic (16:1) | Stearic (18:0) | Oleic (18:1) | Linoleic (18:2) |
| Pro | 27.7 | 3.0 | 40.9 | 0.7 | 15.1 | 36.9 | 3.5 |
| FMAA | 29.5 | 2.2 | 44.6 | 0.8 | 16.3 | 34.4 | 1.7 |
| MIAA | 27.7 | 3.1 | 43.0 | 0.6 | 16.3 | 35.6 | 1.4 |
| SVAA | 23.7 | 2.1 | 46.1 | 0.8 | 14.7 | 35.0 | 1.4 |
| Glucose [5] | 67.5 | 1.3 | 20.0 | 0.6 | 14.6 | 46.9 | 13.1 |
| Glycerol [26] | 35.0 | 1.4 | 27.8 | 0.6 | 21.8 | 43.8 | 2.9 |
| Corn stover [33] | - | 2.6 | 44.6 | 1.0 | 15.8 | 36.0 | 0.7 |
| Sugarcane juice [34] | 45.0 | 1.0 | 21.5 | 0.7 | 4.6 | 62.1 | 7.6 |
| Palm [35] | - | - | 42.7 | - | 2.1 | 38.4 | 10.6 |
| Canola [35] | - | - | 3.7 | 0.2 | 1.9 | 62.4 | 20.1 |

## 4. Conclusions

Here, we showed that the oleaginous yeast *R. toruloides* CGMCC 2.1389 can use most of the 20 proteinogenic AAs individually or in blends with similar AA compositional profiles to those of meat wastes for lipid production according to a two-stage culture mode. The lipids produced from AA herein showed similar fatty acid compositional profiles to those of microbial lipids produced in sugars and related organic substances. Our results suggests that AA wastes can be used as substrates for lipid production, yet this new route requires further investigation to improve the overall efficiency through the identification of cost-effective protein wastes, more robust oleaginous yeast strains, and advanced bioprocesses.

**Supplementary Materials:** The following are available online at http://www.mdpi.com/1996-1073/13/7/1576/s1: Table S1: Gradient condition for analyzing all 20 proteinogenic amino acids with IC. Table S2: Analysis of variances for cell mass in Figure 1 (all AAs). Table S3: Analysis of variances for cell mass in Figure 1 (Gly and Tyr). Table S4: Analysis of variances for cell mass in Figure 1 (Met, His, Arg, Thr, Trp, Tyr, Lys, and Leu). Table S5: Analysis of variances in Figure 2. Figure S1: Chromatogram of standard mixture of all 20 proteinogenic amino acids.

**Author Contributions:** Z.K.Z. conceived the project. Q.L., R.K., and X.Y. designed and performed the experiments. Q.W., Q.L., and R.K. performed ion chromatography analyses. Q.L., R.K., and Z.K.Z. wrote and revised the manuscript. All authors discussed the results and commented on the manuscript. All authors read and approved the final manuscript.

**Funding:** This work was funded by National Natural Science Foundation of China (No. 51761145014 and 21721004).

**Conflicts of Interest:** The authors declare no conflict of interest.

## References

1. Luque, R.; Lovett, J.C.; Datta, B.; Clancy, J.; Campeloa, J.M.; Romero, A.A. Biodiesel as feasible petrol fuel replacement: A multidisciplinary overview. *Energy Environ. Sci.* **2010**, *3*, 1706–1721. [CrossRef]

2. Liang, M.H.; Jiang, J.G. Advancing oleaginous microorganisms to produce lipid via metabolic engineering technology. *Prog. Lipid Res.* **2013**, *52*, 395–408. [CrossRef]

3. Li, Q.; Du, W.; Liu, D. Perspectives of microbial oils for biodiesel production. *Appl. Microbiol. Biotechnol.* **2016**, *80*, 749–756. [CrossRef] [PubMed]

4. Ratledge, C.; Wynn, J.P. The biochemistry and molecular biology of lipid accumulation in oleaginous microorganisms. *Adv. Appl. Microbiol.* **2002**, *51*, 1–51. [PubMed]

5. Li, Y.; Zhao, Z.K.; Bai, F. High-density cultivation of oleaginous yeast *Rhodosporidium toruloides* Y4 in fed-batch culture. *Enzyme Microb. Technol.* **2007**, *41*, 312–317. [CrossRef]

6. Zhao, X.; Hu, C.; Wu, S.; Shen, H.; Zhao, Z.K. Lipid production by *Rhodosporidium toruloides* Y4 using different substrate feeding strategies. *J. Ind. Microbiol. Biotechnol.* **2011**, *38*, 627–632. [CrossRef] [PubMed]

7. Pan, J.G.; Rhee, J.S. Kinetic and energetic analyses of lipid accumulation in batch culture of *Rhodotorula glutinis*. *J. Ferment. Technol.* **1986**, *64*, 557–560. [CrossRef]

8. Hu, C.; Zhao, X.; Zhao, J.; Wu, S.; Zhao, Z.K. Effects of biomass hydrolysis by-products on oleaginous yeast *Rhodosporidium toruloides*. *Bioresour. Technol.* **2009**, *100*, 4843–4847. [CrossRef]

9. Koutinas, A.A.; Papanikolaou, S. Biodiesel production from microbial oil. In *Handbook of Biofuels Production*; Woodhead Publishing Limited: Cambridge, UK, 2011; pp. 178–198. [CrossRef]

10. Koutinas, A.A.; Chatzifragkou, A.; Kopsahelis, N.; Papanikolaou, S.; Kookos, I.K. Design and techno-economic evaluation of microbial oil production as a renewable resource for biodiesel and oleochemical froduction. *Fuel* **2014**, *116*, 566–577. [CrossRef]

11. Li, S.Y.; Ng, I.S.; Chen, P.T.; Chiang, C.J.; Chao, Y.P. Biorefining of protein waste for production of sustainable fuels and chemicals. *Biotechnol. Biofuels* **2018**, *11*, 256. [CrossRef]

12. De Schouwer, F.; Claes, L.; Vandekerkhove, A.; Verduyckt, J.; De Vos, D.E. Protein-rich biomass waste as a resource for future biorefineries: State of the art, challenges, and opportunities. *ChemSusChem* **2019**, *12*, 1272–1303. [CrossRef]

13. Pelley, J.W. *Elsevier's Integrated Biochemistry*, 1st ed.; Elsevier: Philadelphia, PA, USA, 2007; Chapter 17; pp. 97–104.

14. Zeng, Y.; Bian, D.; Xie, Y.; Jiang, X.; Li, X.; Li, P.; Zhang, Y.; Xie, T. Utilization of food waste hydrolysate for microbial lipid and protein production by *Rhodosporidium toruloides* Y2. *J. Chem. Technol. Biotechnol.* **2017**, *92*, 666–673. [CrossRef]

15. Papanikolaou, S.; Aggelis, G. Lipids of oleaginous yeasts. Part I: Biochemistry of single cell oil production. *Eur. J. Lipid Sci. Technol.* **2011**, *113*, 1031–1051. [CrossRef]

16. Wu, S.; Hu, C.; Jin, G.; Zhao, X.; Zhao, Z.K. Phosphate-limitation mediated lipid production by *Rhodosporidium toruloides*. *Bioresour. Technol.* **2010**, *101*, 6124–6129. [CrossRef] [PubMed]

17. Wang, Y.; Zhang, S.; Zhu, Z.; Shen, H.; Lin, X.; Jin, X.; Jiao, X.; Zhao, Z.K. Systems analysis of phosphate-limitation-induced lipid accumulation by the oleaginous yeast *Rhodosporidium toruloides*. *Biotechnol. Biofuels* **2018**, *11*, 148. [CrossRef] [PubMed]

18.  Wu, S.; Hu, C.; Zhao, X.; Zhao, Z.K. Production of lipid from N-acetylglucosamine by *Cryptococcus curvatus*. *Eur. J. Lipid Sci. Technol.* **2010**, *112*, 727–733. [CrossRef]

19.  Toldrá, F.; Mora, L.; Reig, M. New insights into meat by-product utilization. *Meat Sci.* **2016**, *120*, 54–59. [CrossRef]

20.  Jayathilakan, K.; Sultana, K.; Radhakrishna, K.; Bawa, A.S. Utilization of byproducts and waste materials from meat, poultry and fish processing industries: A review. *J. Food Sci. Technol.* **2012**, *49*, 278–293. [CrossRef]

21.  Bhaskar, N.; Modi, V.K.; Govindaraju, K.; Radha, C.; Lalitha, R.G. Utilization of meat industry by broducts: Protein hydrolysate from sheep visceral mass. *Bioresour. Technol.* **2007**, *98*, 388–394. [CrossRef]

22.  Salminen, E.; Rintala, J. Anaerobic digestion of organic solid poultry slaughterhouse waste-a review. *Bioresour. Technol.* **2002**, *83*, 13–26. [CrossRef]

23.  Russ, W.; Meyer-Pittroff, R. Utilizing waste products from the food production and processing industries. *Crit. Rev. Food Sci. Nutr.* **2004**, *44*, 57–62. [CrossRef] [PubMed]

24.  Shen, Q.; Guo, R.; Dai, Z.; Zhang, Y. Investigation of enzymatic hydrolysis conditions on the properties of protein hydrolysate from fish muscle (*Collichthys niveatus*) and evaluation of its functional properties. *J. Agric. Food Chem.* **2012**, *60*, 5192–5198. [CrossRef] [PubMed]

25.  Webster, J.D.; Ledward, D.A.; Lawrie, R.A. Protein hydrolysates from meat industry by-products. *Meat Sci.* **1982**, *7*, 147–157. [CrossRef]

26.  Yang, X.; Jin, G.; Gong, Z.; Shen, H.; Bai, F.; Zhao, Z.K. Recycling biodiesel-derived glycerol by the oleaginous yeast *Rhodosporidium toruloides* Y4 through the two-stage lipid production process. *Biochem. Eng. J.* **2014**, *91*, 86–91. [CrossRef]

27.  Lin, J.; Shen, H.; Tan, H.; Zhao, X.; Wu, S.; Hu, C.; Zhao, Z.K. Lipid production by *Lipomyces starkeyi* cells in glucose solution without auxiliary nutrients. *J. Biotechnol.* **2011**, *152*, 184–188. [CrossRef]

28.  Pelley, J.W. Amino Acid and Heme Metabolism. In *Elsevier's Integrated Biochemistry*; Elsevier: Philadelphia, PA, USA, 2007; pp. 97–105. [CrossRef]

29.  Owen, O.E.; Kalhan, S.C.; Hanson, R.W. The key role of anaplerosis and cataplerosis for citric acid cycle function. *J. Biol. Chem.* **2002**, *277*, 30409–30412. [CrossRef]

30.  Jiao, X.; Zhang, Y.; Liu, X.; Zhang, Q.; Zhang, S.; Zhao, Z.K. Developing a CRISPR/Cas9 system for genome editing in the basidiomycetous yeast *Rhodosporidium toruloides*. *Biotechnol. J.* **2019**, *14*, 1900036. [CrossRef]

31.  Turcotte, G.; Kosaric, N. The effect of C/N ratio on lipid production by *Rhodosporidium toruloides* ATCC 10788. *Biotechnol. Lett.* **1989**, *11*, 637–642. [CrossRef]

32.  Ruiz, S.J.; Van Klooster, J.S.; Bianchi, F.; Poolman, B. Growth inhibition by amino acids in *Saccharomyces cerevisiae*. *bioRxiv* **2017**. [CrossRef]

33.  Dai, X.; Shen, H.; Li, Q.; Rasool, K.; Wang, Q.; Yu, X.; Wang, L.; Bao, J.; Yu, D.; Zhao, Z.K. Microbial lipid production from corn stover by the oleaginous yeast *Rhodosporidium toruloides* using the PreSSLP process. *Energies* **2019**, *12*, 1053. [CrossRef]

34.  Soccol, C.R.; Neto, C.J.D.; Soccol, V.T.; Sydney, E.B.; da Costa, E.S.F.; Medeiros, A.B.P.; Vandenbergh, L.P.S. Pilot scale biodiesel production from microbial oil of *Rhodosporidium toruloides* DEBB 5533 using sugarcane juice: Performance in diesel engine and preliminary economic study. *Bioresour. Technol.* **2017**, *223*, 259–268. [CrossRef] [PubMed]

35.  Zambiazi, R.C.; Przybylski, R.; Zambiazi, M.W.; Menonca, C.B. Fatty acid composition of vegetable oils and fats. *Bol. Do Cent. Pesqui. Process. Aliment.* **2007**, *25*, 111–120. [CrossRef]

36.  Patel, A.; Arora, N.; Sartaj, K.; Pruthi, V.; Pruthi, P.A. Sustainable biodiesel production from oleaginous yeasts utilizing hydrolysates of various non-edible lignocellulosic biomasses. *Renew. Sustain. Energy Rev.* **2016**, *62*, 836–855. [CrossRef]

 *energies*

*Article*

# Lignocellulosic Ethanol in a Greenhouse Gas Emission Reduction Obligation System—A Case Study of Swedish Sawdust Based-Ethanol Production

**Sylvia Haus, Lovisa Björnsson and Pål Börjesson ***

Environmental and Energy Systems Studies, Department of Technology and Society, Lund University, P.O. Box 118, SE-221 00 Lund, Sweden; sylvia.haus@miljo.lth.se (S.H.); lovisa.bjornsson@miljo.lth.se (L.B.)
* Correspondence: pal.borjesson@miljo.lth.se; Tel.: +46-46-2228642

Received: 21 January 2020; Accepted: 23 February 2020; Published: 26 February 2020

**Abstract:** A greenhouse gas (GHG) emission reduction obligation system has been implemented in the Swedish road transport sector to promote the use of biofuels. For transportation fuel suppliers to fulfil this obligation, the volume of biofuel required decreases with decreasing life cycle GHG emission for the biofuel, linking lower GHG emission to higher economic value. The aim of this study was to investigate how the economic competitiveness of a Swedish emerging lignocellulosic-based ethanol production system would be influenced by the reduction obligation. The life cycle GHG emission for sawdust-based ethanol was calculated by applying the method advocated in the EU Renewable Energy Directive (RED II). The saving in GHG emissions, compared with fossil liquid transportation fuels, was 93% for a potential commercial production system in southern Sweden. This, in turn, will increase the competitiveness of sawdust-based ethanol compared to the mainly crop-based ethanol currently used in the Swedish biofuel system, which has an average GHG emission saving of 68%, and will allow for an almost 40% higher price of sawdust-based ethanol, compared to the current price of ethanol at point of import. In a future developed, large-scale market of advanced ethanol, today's GHG emission reduction obligation system in Sweden seems to afford sufficient economic advantage to make lignocellulosic ethanol economically viable. However, in a short-term perspective, emerging lignocellulosic-based ethanol production systems are burdened with economic risks and therefore need additional economic incentives to make a market introduction possible.

**Keywords:** ethanol; lignocellulosic biomass; life cycle assessment; GHG emissions; political incentives; economic performance

---

## 1. Introduction

In 2018, the Swedish Government imposed an obligation on the road transportation sector to reduce the greenhouse gas (GHG) emissions from fossil fuels. This reduction obligation means that fuel suppliers in Sweden are required to blend biofuels into fossil fuels to achieve an overall reduction in GHG emission for the fuel blend compared to a fossil fuel comparator. The target for the reduction obligation is increased annually following a predetermined GHG reduction trajectory. The lower the GHG emission of the biofuel, the lower the amount needed to achieve the required overall reduction in emission. This creates an economic advantage for biofuels with low life cycle GHG emissions. An apparent question is, therefore, whether this increased economic advantage is a sufficiently effective incentive to promote the large-scale commercial production of emerging advanced biofuels with low GHG emissions.

The overall objective of this study was to evaluate the economic consequences of GHG emissions from lignocellulosic ethanol in a reduction obligation system. The life cycle assessment-based calculation method defined in the EU renewable energy directive (RED II) was applied as this calculation method is required in the Swedish reduction obligation system, to the case of ethanol produced from sawdust in a potential commercial ethanol plant in Sweden. Sawdust is seen as a promising feedstock for the production of liquid biofuels in Sweden, compared with different lignocellulosic feedstocks, due to its physical proporties, e.g., a homogenus feedstock with no, or low, impurities, low costs, and abundant volymes in the sawmill sector [1]. The risk for increased GHG emissions due to changes in various factors during the planning process, or after the start of operation, such as production system design, selection of inputs in the process, availability of feedstock, or interpretation of the GHG calculation methodology, are evaluated. The resulting life cycle GHG emission for sawdust-based ethanol is compared with statistics on the average cost and GHG emissions of the ethanol currently used to achieve the Swedish reduction obligation target. Finally, the results are discussed in a broader perspective including previous studies of economy and GHG performance of various lignocellulosic-based ethanol production systems.

## 2. Background

### 2.1. Biofuel Policy in the EU and Sweden

EU member states must require fuel suppliers to supply at least 14% renewable fuels in road and rail transport by 2030, where the amount of advanced biofuels should correspond to 3.5% (percentage points) [2]. The average EU energy use for road transport during the period 2006–2017 was 12,500 PJ year$^{-1}$, with no sign of decline [3]. The average share of renewable fuels (including the double counting allowed for some fuels) was 7.1% in 2016 [4]. An increase to 14% means that the market demand for renewable transport fuels within the EU will approximately double in the coming decade (excluding potential consequences of double counting). All biofuels must fulfil the sustainability criteria set out in RED II, including a GHG emission saving of at least 65% compared to the fossil fuel comparator of 94 g $CO_2$-equivalents ($CO_{2\text{-eq}}$)/MJ, which means a maximum allowed GHG emission of 33 g $CO_{2\text{-eq}}$/MJ [2]. The specified method of calculating biofuel GHG emission is based on an life cycle assessment (LCA) methodology with standardized procedures for system boundaries, functional unit and allocation (see Section 4.1. for a detailed description of this calculation method according to RED II).

Sweden had the highest share of biofuels in domestic road transport in the EU in 2018, with 23% on energy basis (30% including the double counting allowed for some fuels) [5]. The national target set for 2030 is a 70% reduction in GHG emission from domestic transport, compared to 2010 levels. One tool used to bring about this transition is the Swedish GHG emission reduction obligation in road transport, introduced in 2018, which requires fuel suppliers to reduce the GHG emission of petrol and diesel by blending it with biofuels [6]. The calculated reduction in GHG emissions is based on the volume of biofuel utilized in combination with the life cycle GHG emission of the biofuel. Under the GHG reduction obligation system, suppliers of petrol and diesel will need a lower amount of biofuel if it has a low life cycle GHG emission, to achieve an equivalent total emission reduction, compared with a biofuel with a higher life cycle GHG emission. The reduction obligation targets for 2020 are 4.2% for petrol and 21% for diesel (compared to 2010). The reduction levels suggested by the Swedish Energy Agency for 2030 (not yet adopted by the Swedish Parliament) are 28% for petrol and 66% for diesel [6].

### 2.2. Current Production and Use of Fuel Ethanol in the EU and Sweden

In Sweden, 5.3% (vol) ethanol was used for low blend in petrol in 2018, corresponding to 3.2 PJ. For the reduction obligation to be achieved in 2030, the estimated amount of ethanol required is roughly twice that, i.e., 6.8 PJ [7]. During the period from 2011–2018, over 95% of the ethanol used for domestic transport was produced from agricultural crops, and was mainly imported from other EU countries. During the same period, the average ethanol GHG emission has decreased from 37 g $CO_{2\text{-eq}}$/MJ to 30 g $CO_{2\text{-eq}}$/MJ. To achieve the target GHG emission reduction of 70% in the transport sector by 2030, the expected average emission from ethanol is 15 g $CO_{2\text{-eq}}$/MJ [6]. Thus, a shift towards ethanol with lower life cycle GHG emissions is necessary [6].

Default values of ethanol GHG emissions in the RED II for plants using process heat from biomass fuels are 30–31 g $CO_{2\text{-eq}}$/MJ when using crops such as corn and other cereals as ethanol feedstock, and 16 $CO_{2\text{-eq}}$/MJ when wheat straw is used [2]. Straw is an example of a feedstock for so-called "advanced biofuels", as defined in the EU RED II Annex IX, including lignocellulosic waste and residues from forestry and forest-based industries, such as sawdust [2]. The main reason why ethanol based on lignocellulosic residues is calculated to have lower life cycle GHG emissions than ethanol based on cereals, is due to lower GHG emissions from the production of the biomass feedstock (see e.g., [8]). All GHG emissions from the cultivation phase of cerelas are included in the GHG calculation of crop-based ethanol, whereas only the GHG emissions from the recovery of the lignocellulosic residues, and not from the up-stream primary biomass production, are included regarding lignocellulosic residue-based ethanol. The European fuel ethanol use in 2017 was 115 PJ, of which almost 90% was produced within the EU [9]. The GHG emission from EU-produced ethanol, most of which is crop-based, decreased from 42 to 24 g $CO_{2\text{-eq}}$/MJ between 2011 and 2018 [9]. This decrease is not the result of a shift from crops to advanced feedstock, but mainly due to measures permitted in the RED II calculation method, the allocation of emissions to by-products (animal feed), and increasing capture and use of the biogenic $CO_2$ produced during fermentation, allowing this to be subtracted from the life cycle GHG emissions [2,9]. Advanced ethanol represented less than 4% of the total European ethanol production in 2018 [9].

### 2.3. Advanced Ethanol Production

The link between GHG emissions and the economic value of a biofuel created through the reduction obligation system will afford an economic advantage to low-emission biofuels [6]. Lignocellulosic ethanol typically has low life cycle GHG emissions, from the default value of 16 g $CO_{2\text{-eq}}$/MJ for straw given in RED II, to around 10 g $CO_{2\text{-eq}}$/MJ for sawdust-based ethanol production, recently demonstrated on pilot scale in Finland [2,10].

Sweden has a long-standing tradition of research into lignocellulose-based ethanol production [11–13], but has not yet a commercial large-scale plant dedicated to the production of advanced ethanol as transportation fuel. A large proportion of the fuel production cost of a crop-based biofuel is the cost of the biomass feedstock, while the production of advanced biofuels requires installations with higher investment costs [6,14]. Globally, advanced ethanol production has stagnated, mainly due to technical difficulties and high production costs [15], and advanced ethanol made up less than 2% of the ethanol used in Sweden in 2018 [5]. The considerable financial risk means that long-term stable political incentives will be required for this type of ethanol production to be commercialized [6]. The current GHG reduction obligation system in Sweden may be such a long-term stable incentive, promoting the production and use of biofuels with low GHG emissions, and having predetermined reduction targets until 2030.

## 3. The Case Study: Ethanol from Sawdust

To examine the economic advantage of advanced ethanol production in the Swedish reduction obligation system, the case of ethanol production from sawdust was investigated. This case is based on recent data provided in the scientific literature [16,17] and by construction planners for commercial ethanol plants in Sweden with the capacity to process 200,000 ton (dry matter, DM) of sawdust per year [18].

### 3.1. Raw Material Availability—Sawdust from Sawmills in a Forest Dense Region

The current production of sawdust in Swedish sawmills is approximately 1.9 million ton DM per year, which in terms of energy represents some 35 PJ [19]. The region with the highest density of forest, in form of conifers, and sawmills, is found in southern Sweden between latitude 56.5–58 °N and longitude 13.5–16.5 °E. This region amounts to 2.1 million hectares of which 70% is covered by forest [20]. The annual forest increment is equivalent to 14 million $m^3$, while annual felling amounts to 12 million $m^3$ [20]. There are currently 34 sawmills in operation in the region, producing approximately

4 million m$^3$ sawn timber per year, which is equivalent to roughly 8.7 million m$^3$ of roundwood [19]. For comparison, this represents 22% of the total sawn timber production in Sweden.

Based on the generation of, on average, 48 kg sawdust (DM) per m$^3$ of roundwood [21,22], the total annual production of sawdust in sawmills in this forest region is approximately 420,000 ton DM, equivalent to roughly twice the amount needed as feedstock for one ethanol plant of the size assumed in this study.

The theoretical transportation distance of sawdust from existing sawmills in the studied forest region to an ethanol plant with an optimal localization is calculated to be 56 km. This calculation is based on the following assumptions: (i) the supply of sawdust is evenly distributed over the region (based on the actual location of the 34 sawmills), (ii) the sawdust recovery area is circular and the ethanol plant is located at the centre, and (iii) the practical transportation distance is 20% longer than the theoretical linear distance [23]. It was also assumed that all the sawdust produced would be allocated to the ethanol plant. However, only part of the sawdust generated will be commercially available as feedstock for the ethanol plant since sawdust is also used for other purposes, such as the production of pellets, district heating, etc. In the base case in the current study, it is assumed that two thirds of the sawdust produced in the region will be available as feedstock for the ethanol plant, giving an average one-way transportation distance of 70 km. If, on the other hand, only one third of the sawdust is available, the transportation distance will increase to 100 km, which is applied in one of the alternative assessments.

## 3.2. The Ethanol Plant

The process design was determined by data from [16–18,24]. The annual sawdust input and product outputs of the ethanol production plant are illustrated in Figure 1. Detailed information on inputs of material and utilities is presented in Appendix A.

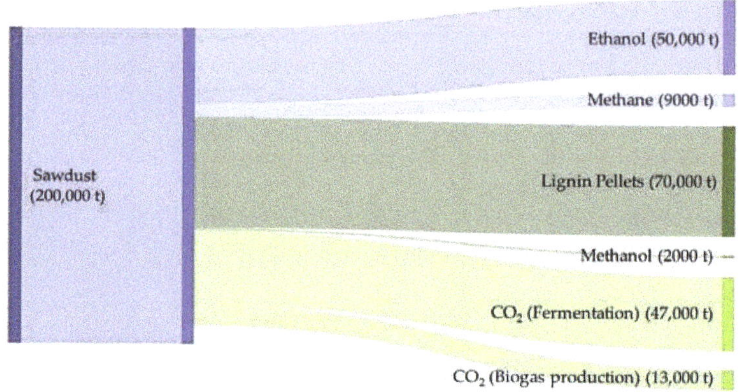


**Figure 1.** Annual feedstock input and ethanol and co-product outputs (DM) at the studied ethanol production plant [ton/year].

The process for lignocellulosic ethanol production is based on previously published research on pilot and demonstration scale (see e.g., [16,17,25]) and is briefly summarized here. The process consists of acid-catalysed steam pre-treatment followed by liquid/solid phase separation. The cellulose- and lignin-containing solid phase is treated by simultaneous saccharification (hydrolysis) and fermentation (SSF). The ethanol produced is recovered through distillation, and the remaining lignin-rich solids are dried and pelletized. The liquid phase from pre-treatment is treated by anaerobic digestion, where the biogas containing methane and carbon dioxide is recovered in the gas phase; further upgrading is not included.

### 3.3. Alternative Cases

We considered six different alternative cases in the case study, in addition to the base case, which was based on the data presented in Section 3.2. The impact on ethanol GHG emissions of factors that could change during the planning process, as a consequence of a change in feedstock availability, or that are the result of alternatives in the calculation method, were evaluated. The alternative cases (A–F) are summarized in Table 1, and explained below.

**Table 1.** Overview of the alternative cases studied.

| Factor | Base Case | Alternative | Change |
|---|---|---|---|
| Enzyme | Best available technology | A | Published data on commercial enzymes |
| Power production | Swedish electricity mix | B | Nordic electricity mix |
| Heat production | Stand-alone wood chip boiler | C | Internal use of lignin pellets |
| Feedstock availability | 2/3 (70 km transport distance) | D | 1/3 (100 km transport distance) |
| Transportation fuel | Swedish diesel blend | E | Biodiesel, HVO100 |
| $CO_2$ capture | No capture | F | 30% capture and use |

Alternative A: The enzyme cocktails added for cellulose hydrolysis have been reported to contribute with a large share of the GHG emissions for wood based ethanol. These emissions are both linked to the required enzyme dose and an impact of the carbon source and energy use during enzyme production [26–28]. For the base case, data representing low emissions for a future European cellulose enzyme production are chosen based on [27], and an enzyme dose in the lower range (0.4 g enzyme protein per MJ ethanol produced) suggested in the same study, corresponding to 2.7 kg enzyme protein per t DM sawdust added. The emission data are based on enzyme production with sugar beet molasses as carbon source, and with biogas as energy source for electricity, heating and cooling, giving a carbon footprint of 6 kg $CO_{2\text{-eq}}$/kg enzyme protein [27]. As an alternative, published data for commercially available enzymes, with an enzyme dose corresponding to 30.4 kg enzyme formula per t DM sawdust added, and a carbon footprint of 5.5 kg $CO_{2\text{-eq}}$/kg enzyme formula are evaluated [29] (N.B. Alternative A data are given in weight enzyme formula, while the base case data are given per weight enzyme protein).

Alternative B: In the base case, the carbon footprint of the electricity mix is based on Swedish national emissions, in line with the RED II methodology [2,30,31]. To illustrate the impact of higher GHG emissions for electricity, the Nordic electricity mix was evaluated [32].

Alternative C: The fuel for the generation of process steam is assumed to be wood chips in the base case. The largest quantity of product from the ethanol plant is lignin pellets (Figure 1). Therefore, the use of internally produced lignin pellets for steam production is evaluated.

Alternative D: The proportion of sawdust in the province available to the ethanol plant is decreased from two thirds to one third, increasing the one-way transport distance from 70 km to 100 km, as described in Section 3.1.

Alternative E: The impact of changing the type of transportation fuel from the current Swedish average diesel blend (77% fossil diesel and 23% biodiesel) to biodiesel (Hydrotreated Vegetable Oil, HVO100) is evaluated [5].

Alternative F: The average share of $CO_2$ capture from fermentation in European ethanol production was 18% in 2018 [9]. The possibility of $CO_2$ capture was evaluated, assuming that 30% of the $CO_2$ from the fermentation process (equal to 14,100 t/y, Figure 1) is recovered for further use, replacing fossil $CO_2$.

## 4. Methods and Data Inventory

### 4.1. Revised Renewable Energy Directive (RED II)

As described in previous sections, the life cycle-based GHG emission calculations were performed according to the revised European Union's Renewable Energy Directive, RED II [2], which is to be implemented in national legislation after 2020. The reason why this specific methodology was utilized is that the Swedish GHG reduction obligation system is built upon the GHG performance of the blend in biofuels according to the calculation method defined in the EU RED.

According to the RED II [2], the GHG performance was calculated as global warming potential with a 100-year time frame including emissions of $CO_2$, $CH_4$ and $N_2O$, were 1 g of $CH_4$ and 1 g $N_2O$ were valued 25 and 298 g $CO_{2\text{-eq}}$, respectively. The emissions of the intermediate and final products were expressed in g $CO_{2\text{-eq}}$, and the functional unit (FU) to which environmental impact is related was 1 MJ (lower heating value, LHV) of ethanol. According to the RED II, the default value for "the fossil fuel comparator $E_{F(t)}$ was 94 g $CO_{2\text{-eq}}$/MJ" (petrol and diesel) [2]. The equation used to calculate the total life cycle emission from the produced fuel is given in RED II [2] as follows:

$$\text{"}E = e_{ec} + e_l + e_p + e_{td} + e_u - e_{sca} - e_{ccs} - e_{ccr},\text{"} \tag{1}$$

where "$E$ is the total emissions from the use of the fuel, $e_{ec}$ is emissions from the extraction or cultivation of raw materials, $e_l$ is annualised emissions from carbon stock changes caused by land-use change, $e_p$ is emissions from processing, $e_{td}$ is emissions from transport and distribution, $e_u$ is emissions from the fuel in use, $e_{sca}$ is emission savings from soil carbon accumulation via improved agricultural management, $e_{ccs}$, is emission savings from $CO_2$ capture and geological storage, $e_{ccr}$ is emission savings from $CO_2$ capture and replacement".

In the present assessment, "emissions from carbon stock changes caused by land-use change" was not relevant since the sawdust was defined as "advanced feedstock" and set to zero, according to RED II [2]. The EU RED also state that the "emissions from the fuel in use shall be taken to be zero for biofuels and bioliquids" [2], and therefore were not included. Furthermore, "emission savings from soil carbon accumulation via improved agricultural management", as well as "emission savings from $CO_2$ capture and geological storage" [2], were not relevant in this study. "Emission savings from $CO_2$ capture and replacement" [2], were set to be zero in the base case, but were included in Alternative F (the $CO_2$ capture case). These emission savings shall, according to RED II [2], "be related directly to the production of biofuel or bioliquid they are attributed to, and shall be limited to emissions avoided through the capture of $CO_2$ of which the carbon originates from biomass and which is used to replace fossil-derived $CO_2$ in commercial products and services". Thus, the parameters in the RED II calculation methodology that were included in following life cycle GHG emission analysis (base case) are "$e_{ec}$", which includes the production of chemicals and enzymes used in the ethanol process (and no up-stream primary biomass production activites since sawdust is seen as a residue), "$e_p$", covering the production of the electricity and heat needed for the ethanol process, and "$e_{td}$" which includes the transport operations for the sawdust from sawmills to the ethanol plant, the transport of the ethanol to the depot and the distribution of the ethanol to the filling stations.

*4.2. LCA Data Inventory*

It is assumed that sawdust (50% DM) [33] from the sawmill is transported via road by truck (40 ton total weight) with a load capacity of 26 tons. Data on diesel use during transport and the properties of the truck were based on [34]. The total fuel use for transport, including the empty return of the truck, was calculated to be 0.12 MJ/kg DM sawdust in the base case, for a transport distance of 70 km, and 0.17 MJ/kg DM sawdust in Alternative D where the transport distance was 100 km. The fuel consumption for transporting the ethanol produced 150 km to a depot, and then a further 150 km to a filling station, based on assumed distances in [35], was calculated to be 0.26 MJ/kg ethanol, including the empty return of a truck for liquids. Electricity use at the depot and at the filling station were assumed to be 0.84 kJ electricity/MJ ethanol and 34 kJ electricity/MJ ethanol, respectively [35]. It was assumed that the heat required for steam production in the base case was generated from forest residues (branches and tops) in a stand-alone wood chip boiler with a conversion efficiency of 95%. It was assumed that the electricity was generated externally and was the Swedish electricity mixture.

The contributions of the chemicals and utilities used in the ethanol process to GHG emission are given in Table 2.

**Table 2.** Input of chemicals, nutrients and enzymes in the production process.

| Input | [kg $CO_{2\text{-eq}}$/kg] | Reference |
|---|---|---|
| Sulphur dioxide | 0.36 | [36] |
| Sodium hydroxide | 0.95 | [36] |
| Sulphuric acid | 0.09 | [36] |
| Antifoam | 1.33 | [29] |
| Trace minerals | 0.44 | [37] |
| Urea | 2.63 | [36] |
| Enzymes: low carbon footprint | 6.05 [a] | [27] |
| Enzymes: Alternative A | 5.50 [a] | [29] |
| | [g $CO_{2\text{-eq}}$/MJ] | |
| Swedish electricity mix | 13.1 | [30] |
| Nordic electricity mix | 34.9 | [32] |
| Heat (wood chip boiler) | 3.4 [b] | [38] |
| Diesel (77% diesel/23% biodiesel) | 77.2 | [5] |
| HVO100 | 8.8 | [5] |

[a] The emissions from enzyme production in the base case is based on future production data with renewable energy and sugar beet molasses as carbon source, and is given per kg enzyme protein. The Alternative A data are given for the commercial product Cellic®Ctec 3 (Novozymes) and per kg enzyme formula; [b] Per MJ wood chips.

In Alternative C, where a fraction of the lignin pellets is used for internal steam production, 47% of the produced lignin pellets was required for heat production, leaving 805 TJ lignin pellets per year as a co-product. This change in the amount of co-product changes the proportion of GHG emissions allocated to ethanol, according to the RED II calculation methodology. In Alternative F, an additional 134 kWh/t $CO_2$ recovered was assumed to be needed for the compression of $CO_2$ [39].

### 4.3. The Swedish GHG Reduction Obligation System and Economic Background Data

The life cycle GHG emissions calculated in the current study were used as input values for the economic assessment of sawdust-based ethanol. Results are presented for the base case, and for the alternatives that give the highest and lowest emissions.

The fuel price at the filling station when the fossil fuel supplier complies with the reduction obligation by blending ethanol with current GHG emission, and price is given in Table 3. The reduction obligation for 2020 of 4.2% was used [6], allowing an emission of 89.4 g $CO_{2\text{-eq}}$/MJ for petrol. With the current average ethanol GHG emission, this required a blend-in of 9.2% (vol) ethanol, giving a petrol price of 19.7 € $GJ^{-1}$ (excl. taxes), based on the data in Table 3. This price per energy unit was kept constant, and the lower volume needed for ethanol with lower GHG emissions was recalculated to give an ethanol price at the filling station. The cost for domestic storage and distribution (including labour and capital costs) and net margins (including profit) were taken from an inventory by the Swedish Energy Agency for E85 in 2018 [40], and were assumed to be the same for sawdust-based ethanol. The values used and the references are presented in Table 3.

**Table 3.** Data used for economic calculations in the production of sawdust-based ethanol.

| Parameter | Unit | Value | Reference |
|---|---|---|---|
| Fossil petrol comparator | g CO$_{2\text{-eq}}$/MJ | 93.3 | [41] |
| LHV petrol [a] | MJ/L | 32.2 | [42] |
| Tax (energy and CO$_2$) [b] | €/L | 0.64 | [43] |
| Petrol price [c] | €/L | 0.59 | [44] |
| Current GHG emission from ethanol (2018) [d] | g CO$_{2\text{-eq}}$/MJ | 30 | [6] |
| Current ethanol price at filling station (excl. taxes) [c] | €/L | 0.896 | [44] |
| Cost of domestic storage and distribution of ethanol [e] | €/L | 0.017 | [40] |
| Cost ethanol net margin (including profit) | €/L | 0.372 | [40] |
| Exchange rate | SEK/€ | 10.33 | [43] |

[a] Petrol typically used in Sweden differs somewhat from the European average (32.0 MJ/L); [b] Tax levied on both petrol and ethanol within the reduction obligation since 1 July 2019. The current tax is per L fuel, not per energy unit, which gives the impact that renewable ethanol is taxed higher than fossil petrol per energy unit; [c] Average price July-October 2019 to costumer at manned filling stations, excluding Value Added Tax (VAT) (25%) and tax; [d] Average GHG emission from ethanol used as biofuel in Sweden in 2018; [e] Including labour and capital costs.

## 5. Results and Discussion

### 5.1. Ethanol Production and GHG Performance

A commercial sawdust-based ethanol plant of the scale evaluated in the current study would use roughly half of the sawdust generated in sawmills located in a forest region in southern Sweden having a high sawmill density, and some 10% of the total Swedish sawdust potential. The 63 million L of ethanol produced corresponds to 1.3 PJ, or 19%, of the ethanol demand required to fulfil the proposed reduction obligation in 2030 [7,45]. The theoretical maximum ethanol production potential from sawdust in Sweden is consequently twice the national demand expected in 2030.

The ethanol produced from sawdust will have a life cycle GHG emission of 6.7 g CO$_{2\text{-eq}}$/MJ in the base case (see Figure 2), which is equivalent to a GHG emission reduction of 93% compared to the fossil fuel reference value of 94 g CO$_{2\text{-eq}}$/MJ given in RED II. The emission from sawdust-based ethanol is less than one fourth of the life cycle emission of 30 g CO$_{2\text{-eq}}$/MJ from the ethanol currently used in Sweden.

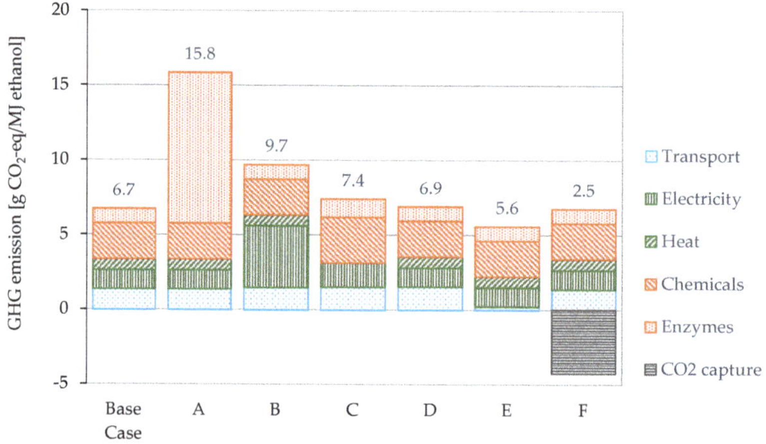


**Figure 2.** GHG emissions [g CO$_{2\text{-eq}}$/MJ Ethanol] for the different alternatives. Net values shown above bars.

The results of the alternative calculations presented in Figure 2 show that the GHG emission could have varied from 2.5 to 15.8 g $CO_{2\text{-eq}}$/MJ, depending on system design and calculation assumptions. For example, changes in the type and dosage of enzymes may have more than doubled the GHG emissions, whereas changes in the electricity mix may have increased the emission by almost 50%. This clearly shows the importance of using enzymes with low carbon footprint in combination with low enzyme dosage, together with electricity having low GHG emissions. On the other hand, capture of biogenic $CO_2$, which was used to replace fossil-derived $CO_2$, may have reduced the life cycle GHG emissions by more than 50%. A change in transportation distance had a minor impact, whereas the replacement of diesel fuel by biodiesel (100% HVO) led to a reduction in GHG emission of approximately 15%. Finally, the use of lignin pellets for the internal generation of process heat, instead of an external supply of heat based on wood chips, led to a somewhat poorer GHG performance (approximately 10% higher life cycle GHG emissions). This is due to the calculation methodology in RED II and the allocation rules regarding co-products.

The results presented in Figure 2 regarding the GHG emissions for sawdust-based ethanol can be compared with previous studies of other lignocellulosic ethanol production systems. Becker et al. [46] analysed the GHG emissions of ethanol production from wood chips from logging residues (tops and branches after final felling), short-rotation coppice willow, and straw, also according to the EU RED calculation method. The GHG emissions presented for logging residue- and straw-based ethanol were 5.4 and 5.2 g $CO_{2\text{-eq}}$/MJ, respectively [46], which was similar to the base case result for sawdust-based ethanol shown in Figure 2. Both logging residues and straw are defined as biomass residues, similar to sawdust, thus no upstream GHG emissions from the primary biomass production (round wood production and cereal cultivation, respectivley) are included. However, when a dedicated energy crop such as short-rotation coppice willow was used as feedstock, the GHG emissions were somewhat higher, 16.1 g $CO_{2\text{-eq}}$/MJ, due to the inclusion of the GHG emission during the cultivation phase [46]. A conclusion is therefore that the economic competitiviness for willow-based ethanol systems will be somewhat reduced in a GHG reduction obligation system, compared with lignocellulosic waste-based ethanol systems (see Section 5.2).

In a study by Lantz et al. [47], the GHG emissions of both straw- and grain-based ethanol were calculated (winter wheat) based on the EU RED calulation method. Their results showed somewhat higher GHG emissions for straw-based ethanol, or 11 g $CO_{2\text{-eq}}$/MJ, which can be explained by higher GHG emissions from the electricity in use which was based on Nordic electricity mix. This system is reflected in Alternative B in Figure 2, which is also based on Nordic electricity mix instead of Swedish electricity mix. The Swedish national regulations of the EU RED about the GHG calculation method were revised in 2018, including a change from the requirement of using Nordic electricity mix to the requirement of using Swedish electricity mix [31].

The RED calculation method applied in EU legislation (Section 4.1) is a simplified LCA approach, based on the ISO standard of LCA [48,49]. Several previous studies assessing the GHG emissions of biomass-based ethanol systems have applied somewhat altered calcualtion methods, e.g., including indirect effects of the production system by expanding the system boundaries, substituion effects from by-products etc (see e.g., [29,38,47,50,51]). Depending on the aim of the study, and the life cycle GHG calculation approach, the results will differ, which can lead to different conclusions when comparing the GHG performance of ethanol production systems. However, since the EU RED calculation method is applied in EU regulations and in national policy instruments, such as the Swedish GHG reduction obligation system, this method will be applied by all actors within the biofuel sector in EU, and will be the basis of comparison of economic consequences for various biofuels.

## 5.2. Economic Consequences of GHG Performance

The GHG emission from the production of sawdust-based ethanol in the base case was used together with the alternatives that gave the highest (A) and lowest (F) GHG emissions in an economic assessment. These values are given in Table 4, together with the average GHG emission for fuel ethanol currently used in Sweden. The blend-in demand required to fulfil the 2020 reduction obligation is also given, together with the calculated prices of sawdust-based ethanol.

**Table 4.** Blend-in demand and price of ethanol in the Swedish reduction obligation system 2020.

| Parameter | Unit | Fuel Ethanol | Sawdust-Based Ethanol | | |
| --- | --- | --- | --- | --- | --- |
| | | Average Sweden, 2018 | High | Base Case | Low |
| Ethanol GHG emission | g $CO_{2\text{-eq}}$/MJ | 30 | 15.8 | 6.7 | 2.5 |
| Ethanol blend-in | % energy | 6.2 | 5.1 | 4.5 | 4.3 |
| | % volume | 9.2 | 7.6 | 6.8 | 6.5 |
| Energy (Lower Heating Value) fuel blend | MJ/L | 31.2 | 31.4 | 31.4 | 31.5 |
| Price of fuel to customer (excl. taxes) [a] | €/GJ | 19.7 | | | |
| Price of fuel to customer (excl. taxes) | €/L | 0.615 | 0.619 | 0.620 | 0.621 |
| Price of ethanol at filling station (incl. distribution and margin, etc.) | €/L | 0.90 | 1.01 | 1.09 | 1.12 |
| Price of ethanol at the point of import | €/L | 0.51 | 0.62 | 0.70 | 0.73 |

[a] Based on current ethanol and petrol price (2019) and current (2018) GHG emissions for fuel ethanol used in Sweden. Used as a reference value for the sawdust-based ethanol calculations.

The price of ethanol at the point of import is shown in Figure 3. The cost of storage, distribution and net margin for sawdust-based ethanol was assumed to be the same as for ethanol currently used in Sweden (2018). The resulting comparable, GHG-adjusted, price of sawdust-based ethanol in the base case was equivalent to 0.70 €/L. In other words, the price of sawdust-based ethanol can theoretically be 37% higher than the current price of crop-based ethanol (0.51 €/L) for the fuel supplier whithout leading to an increased fuel price to the customer, due to the lower GHG emissions for the sawdust-based ethanol (6.7 g $CO_{2\text{-eq}}$/MJ) compared to the currently used crop-based ethanol (30 g $CO_{2\text{-eq}}$/MJ). Process improvement such as $CO_2$ capture (Low, Table 4) would reduce the GHG emission to below 3 g $CO_{2\text{-eq}}$/MJ, and allow for a 44% higher sawdust-based ethanol price. If, on the other hand, the life cycle emission was above 15 g $CO_{2\text{-eq}}$/MJ, exemplified here by increased emissions related to the use of enzymes (High, Table 4), the GHG-adjusted sawdust-based ethanol price would be only 22% higher than that of ethanol currently used in Sweden.

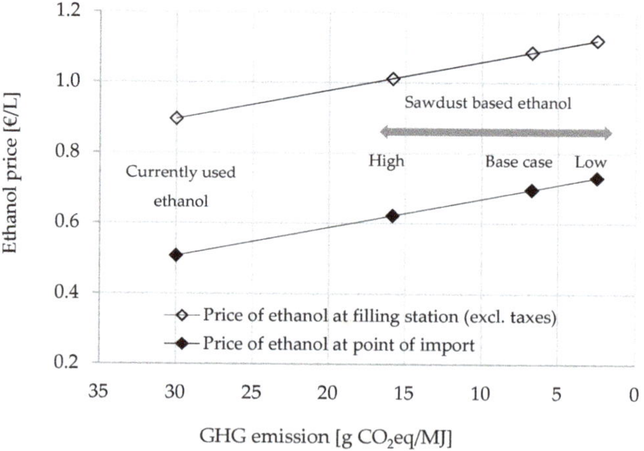


**Figure 3.** The relation between GHG emission and GHG-adjusted ethanol price in the Swedish reduction obligation system based on current fuel prices (2018), and assuming the same average GHG emission as for fuel ethanol used in Sweden in 2018.

The GHG adjusted ethanol price presented in Figure 3 for sawdust-based ethanol can be compared with other lignocellulosic ethanol production systems, which are described in Section 5.1. For example, ethanol based on logging residues and straw has been shown to have similar life cycle GHG emissions as sawdust-based ethanol [46]. Therefore, the price advantage compared to the currently used crop based ethanol would be similar for ethanol from these feedstocks in the Swedish GHG reduction obligation system. Regarding short-rotation coppice willow-based ethanol [46], the GHG emissions are comparable with the alternative "High" in Figure 3, thus equivalent to a possible 22% higher price compared to the ethanol currently used in Sweden.

The sort of findings presented in this paper will be increasingly valuable from the perspective of a commercial operator planning to invest in advanced ethanol production, when new economic policy instruments based on biofuel life cycle GHG emissions are introduced. These findings will also be increasingly valuable for policy makers in designing similar policy tools, thereby striving to promote advanced biofuels with low GHG emissions. The overall aim of this study was to show that the introduction of a GHG emission reduction obligation system will favour lignocellulosic ethanol. The results show that this policy instrument will allow for a significantly higher increase in the price of lignocellulosic ethanol, compared with the price of biofuels currently used with higher GHG emission. The next question would then be, whether the level of this potential price increase is enough to stimulate the large-scale commercial introduction of lignocellulosic ethanol production systems.

Several previous studies have shown that lignocellulosic ethanol production systems require economic support to be competitive with existing crop-based ethanol production systems, and especially so compared to fossil liquid fuels [15,52–54]. The estimated production cost in the large-scale commercial production of lignocellulosic ethanol has been reported to be up to 30% higher than the production cost using existing wheat-based ethanol production in Sweden [47,51]. The production cost of wheat-based ethanol is in the same range as the current price of ethanol at point of import [47,51] (see Figure 3), thus indicating that future sawdust-based ethanol on a developed market could theoretically be competitive under the existing reduction obligation system as long as the life cycle GHG emission is limited to 10 g $CO_{2\text{-eq}}$/MJ or below.

A recent summary of the costs of producing advanced ethanol (i.e., from agricultural residues and woody biomass) found them to be in the range of 0.51–1.2 €/L [15]. In a techno-economic assessment focusing on woody biomass [16], it was concluded that the minimum ethanol selling price, ensuring profitable production, varied between 0.55 and 1.1 €/L ethanol. The lower minimum ethanol selling price refers to the cheapest white wood feedstock (no bark) available today, such as sawdust, while the higher price refers to feedstock with a high fraction of bark (80%), such as hog fuel. The production of ethanol from logging residues was found to have a minimum ethanol selling price of around 0.70–1.1 €/L. Thus, these results indicate that economic viability is possible with sawdust-based ethanol in a future developed market with large-scale production through the reduction obligation system, allowing a price of around 0.70 €/L in the base case, compared to current crop-based ethanol having a selling price of around 0.50 €/L. These results also indicate that ethanol systems based on logging residues need a somewhat higher selling price than sawdust-based ethanol, even though the two systems have similar GHG performance (see Section 5.1.), due to a slightly higher production cost for logging residue-based ethanol.

However, the comparisons above are based on the assumption of a developed commercial market of advanced ethanol including a large-scale production of e.g., sawdust-based ethanol production at several production sites. The production cost is normally higher for the initial commercial production plants implementing an emerging technology. This is due to remaining technological risks, not fully optimised production systems, unforeseen events, etc [51,55]. Thus, investments in emerging technologies requires a risk compensation in form of, for example, higher selling prices, investment subsidies reducing the capital costs, etc. Furthermore, the above comparisons are based on the assumption of an optimised ethanol production system from a GHG perspective, leading to very high GHG savings compared with fossil fuels. As shown in this study, a less optimised sawdust based ethanol production system, with an emission of 15 g $CO_{2\text{-eq}}$/MJ (Figure 3), will give an important contribution to the shift towards biofuels with low GHG emissions required for 2030, but the price advantage created through the reduction obligation system might be too low to promote such a production.

To ensure that a specific lignocellulosic ethanol production system will be sufficiently profitable and competitive in a future commercial market including a reduction obligation system, specific local conditions and actual system designs must be taken into account. For example, the alternatives presented in this paper (see Figure 2) show that the GHG emissions could be both decreased, for example, by $CO_2$ capture and use, or increased, for example, by uncertainties related to the type and dosage of enzymes used. This also apply to other lignocellulosic-based ethanol production system discussed in this paper. As shown in Figure 3, this will affect the GHG-adjusted price of sawdust-based ethanol (as for alternative lignocellulosic-based ethanol). In addition, the current market price of ethanol, which is mainly imported crop-based ethanol, may also change over time, affecting the future economic viability of lignocellulosic ethanol. To conclude, the Swedish reduction obligation system appears to be sufficient to promote the commercial production of primarily sawdust-based ethanol, among various lignocellulosic-based ethanol systems, under future conditions in a large-scale developed market of advanced ethanol. The reduction obligation system is also a long-term political tool with a suggested reduction target for 2030, which is another important prerequisite if investors are to minimize financial risks. However, in a short-term perspective, additional economic incentives are needed for the implementation of these emerging production systems.

## 6. Conclusions

The main conclusions of the study presented in this paper can be summarized as follows.

- Sawdust-based ethanol can be produced with low life cycle GHG emission, leading to a GHG emission saving of 93% compared with fossil liquid transportation fuels, but it may vary between 83% and 97%.

- This, in turn, will increase the economic competitiveness of sawdust-based ethanol in the road transport sector under the Swedish GHG reduction obligation system, which promotes biofuels with low GHG emissions.

- Based on the current price of ethanol at point of import, and estimated future production costs of lignocellulosic ethanol in a large-scale developed market of advanced ethanol, calculations indicate that sawdust-based ethanol could become economically viable, and potentially also other lignocellulosic waste-based ethanol systems.

- However, in a short-term perspective, emerging sawdust-based ethanol production systems, as well as other lignocellulosic-based ethanol systems, are burdened with higher costs and economic risks and therefore need additional economic incentives to make a market introduction possible.

- The current GHG emission reduction obligation system in Sweden is a long-term stable political incentive, and seems to have the potential to promote future investments in lignocellulosic ethanol production systems in a developed, large-scale market.

**Author Contributions:** S.H. performed the life cycle assessment (LCA) of the ethanol production systems and collected most of the data required for this. L.B. performed the economic assessment, including data collection. L.B. also contributed to the LCA regarding data collection and calculations, and collected data for and wrote the background description. P.B. performed the regional and national analysis, and collected the necessary data. P.B. also provided some data used in the background description, LCA and economic assessment. L.B. and P.B. developed the outline of the study. All authors contributed to the writing and editing of the original manuscript. All authors have read and agreed to the published version of the manuscript.

**Funding:** This research was funded by the Swedish Energy Agency, grant number 41251-1.

**Acknowledgments:** The authors gratefully acknowledge SEKAB for their valuable input in the study. The authors also gratefully acknowledge two anonymous reviewers for their constructive comments.

**Conflicts of Interest:** The authors declare no conflicts of interest.

## Appendix A

**Table A1.** Detailed description of material input per year based on data from [16–18,27].

| Utilities | Input | Flow | Unit |
|---|---|---|---|
| Feedstock | Sawdust (DM) | 200,000 | t |
| Chemicals | Sulphur dioxide (liquid) | 2000 | t |
| | Sodium hydroxide (50% wt.) | 6100 | t |
| | Sulphuric acid (50% wt.) | 960 | t |
| | Antifoam | 3200 | t |
| | Trace minerals | 0.2 | t |
| | Urea (40% wt.) | 12 | t |
| | Enzyme protein | 530 | t |

**Table A2.** Detailed description of utilities required per year [16–18].

| Utilities | Input | Flow | Unit |
|---|---|---|---|
| Energy | Electricity | 89,600 | MWh |
| | Steam (16 bar) | 350,000 | t |
| Water | Process water | 883,000 | t |
| | Boiler feed water (3 bar) | 97,000 | t |

**Table A3.** Detailed description of product output per year [16–18].

| Utilities | Output | Flow | Unit |
|---|---|---|---|
| Products | Ethanol | 50,000 | t |
| | Methanol | 1600 | t |
| | Biogas from biogas generation (65% vol. $CH_4$/35% vol. $CO_2$) | 22,000 | t |
| | Dried lignin (10% moisture) | 78,000 | t |
| | $CO_2$ from fermentation | 47,000 | t |


## References

1. Börjesson, P. *Potential för Ökad Tillförsel Och Avsättning av Inhemsk Biomassa i en Växande Svensk Bioekonomi. Report No 97*; Environmental and Energy Systems Studies, Lund University: Lund, Sweden, 2016.
2. European Union Directive (EU) 2018/2001 of the European Parliament and of the Council of 11 December 2018 on the Promotion of the Use of Energy from Renewable Sources. Available online: https://eur-lex.europa.eu/legal-content/EN/TXT/PDF/?uri=CELEX:32018L2001&from=EN2018 (accessed on 11 December 2018).
3. EEA Final Energy Consumption by Mode of Transport. European Energy Agency. Available online: https://www.eea.europa.eu/data-and-maps/indicators/transport-final-energy-consumption-by-mode/assessment-9 (accessed on 4 October 2019).
4. Eurostat Share of Energy from Renewable Sources. Available online: https://ec.europa.eu/eurostat/statistics-explained/index.php/Renewable_energy_statistics (accessed on 7 October 2019).
5. *Drivmedel 2018. ER 2019:14*; Swedish Energy Agency: Estone, Sweden, 2019.
6. Andrén, R.; Westerberg, N. *Kontrollstation 2019 för Reduktionsplikten. Reduktionspliktens Utveckling 2021–2030. Energimyndigheten*; The Swedish Energy Agency: Eskilstuna, Sweden, 2019.
7. *Komplettering Till Kontrollstation 2019 för Reduktionsplikten*; Swedish Energy Agency: Estone, Sweden, 2019.
8. Börjesson, P.; Ahlgren, S.; Berndes, G. The climate benefit of Swedish ethanol: Present and prospective performance. *Wiley Interdiscip. Rev. Energy Environ.* **2012**, *1*, 81–97. [CrossRef]
9. ePURE European Renewable Ethanol—Key Figures 2018. Available online: https://epure.org/media/1920/190828-def-data-statistics-2018-infographic.pdf (accessed on 13 November 2019).
10. St1 Cellunolix®. Available online: https://www.st1.eu/ (accessed on 3 December 2019).
11. Galbe, M.; Zacchi, G. A review of the production of ethanol from softwood. *Appl. Microbiol. Biotechnol.* **2002**, *59*, 618–628. [CrossRef] [PubMed]

12. Hahn-Hagerdal, B.; Karhumaa, K.; Fonseca, C.; Spencer-Martins, I.; Gorwa-Grauslund, M.F. Towards industrial pentose-fermenting yeast strains. *Appl. Microbiol. Biotechnol.* **2007**, *74*, 937–953. [CrossRef] [PubMed]

13. Taherzadeh, M.J.; Karimi, K. Acid-based hydrolysis processes for ethanol from lignocellulosic materials: A review. *Bioresources* **2007**, *2*, 472–499.

14. Kang, Q.; Appels, L.; Tan, T.; Dewil, R. Bioethanol from lignocellulosic biomass: Current findings determine research priorities. *Sci. World J.* **2014**, *2014*, 13. [CrossRef]

15. Padella, M.; O'Connell, A.; Prussi, M. What is still limiting the deployment of cellulosic ethanol? Analysis of the current status of the sector. *Appl. Sci.* **2019**, *9*, 4523. [CrossRef]

16. Frankó, B.; Galbe, M.; Wallberg, O. Bioethanol production from forestry residues: A comparative techno-economic analysis. *Appl. Energy* **2016**, *184*, 727–736. [CrossRef]

17. Joelsson, E.; Wallberg, O.; Börjesson, P. Integration potential, resource efficiency and cost of forest-fuel-based biorefineries. *Comput. Chem. Eng.* **2015**, *82*, 240–258. [CrossRef]

18. Alwarsdotter, Y. *Personal Communication*; SEKAB: Örnsköldsvik, Sweden, 2019.

19. Swedish Forestry Industries Skogsindustrierna. Available online: https://www.skogsindustrierna.se (accessed on 30 November 2019).

20. Swedish Forest Agency Skogsstyrelsen—Statistik. Available online: https://www.skogsstyrelsen.se/statistik/ (accessed on 30 November 2019).

21. *Sågverk, Branschfakta Nr 1, Stockholm*; Swedish Environmental Protection Agency: Stockholm, Sweden, 2010.

22. Ringman, M. *Trädbränslesortiment: Definitioner Och Egenskaper*; Sveriges Lantbruksuniversitet, Institutionen för Virkeslära: Uppsala, Sweden, 1996.

23. Börjesson, P.; Gustavsson, L. Regional production and utilization of biomass in Sweden. *Energy* **1996**, *21*, 747–764. [CrossRef]

24. Joelsson, J.; Di Fulvio, F.; De La Fuente, T.; Bergström, D.; Athanassiadis, D. Integrated supply of stemwood and residual biomass to forest-based biorefineries. *Int. J. For. Eng.* **2016**, *27*, 115–138. [CrossRef]

25. Sassner, P.; Galbe, M.; Zacchi, G. Techno-economic evaluation of bioethanol production from three different lignocellulosic materials. *Biomass Bioenergy* **2008**, *32*, 422–430. [CrossRef]

26. Dunn, J.B.; Mueller, S.; Wang, M.; Han, J. Energy consumption and greenhouse gas emissions from enzyme and yeast manufacture for corn and cellulosic ethanol production. *Biotechnol. Lett.* **2012**, *34*, 2259–2263. [CrossRef] [PubMed]

27. Gilpin, G.S.; Andrae, A.S. Comparative attributional life cycle assessment of European cellulase enzyme production for use in second-generation lignocellulosic bioethanol production. *Int. J. Life Cycle Assess.* **2017**, *22*, 1034–1053. [CrossRef]

28. Slade, R.; Bauen, A.; Shah, N. The greenhouse gas emissions performance of cellulosic ethanol supply chains in Europe. *Biotechnol. Biofuels* **2009**, *2*, 15. [CrossRef] [PubMed]

29. Olofsson, J.; Barta, Z.; Börjesson, P.; Wallberg, O. Integrating enzyme fermentation in lignocellulosic ethanol production: Life-cycle assessment and techno-economic analysis. *Biotechnol. Biofuels* **2017**, *10*, 51. [CrossRef]

30. Moro, A.; Lonza, L. Electricity carbon intensity in European Member States: Impacts on GHG emissions of electric vehicles. *Transp. Res. Part D Transp. Environ.* **2018**, *64*, 5–14. [CrossRef]

31. Swedish Energy Agency Nya Regler för Hållbarhetskriterier för Biodrivmedel Och Flytande Biobärnslen. Available online: https://www.energimyndigheten.se/ (accessed on 6 February 2020).

32. *Vägledning Till Regelverket Om Hållbarhetskriterier för Biodrivmedel Och Flytande Biobränslen. ER 2012:27*; Swedish Energy Agency: Estone, Sweden, 2012.

33. Eriksson, E.; Gillespie, A.R.; Gustavsson, L.; Langvall, O.; Olsson, M.; Sathre, R.; Stendahl, J. Integrated carbon analysis of forest management practices and wood substitution. *Can. J. For. Res.* **2007**, *37*, 671–681. [CrossRef]

34. Volvo Emissions from Volvo's Trucks. Available online: www.volvotrucks.com (accessed on 31 October 2019).

35. BioGrace BioGrace Version 4d. Available online: https://www.biograce.net/content/ghgcalculationtools/standardvalues (accessed on 30 November 2019).

36. Ecoinvent Ecoinvent 3.5. Available online: https://www.ecoinvent.org/home.html (accessed on 26 November 2019).

37. Dunn, J.B.; Adom, F.; Sather, N.; Han, J.; Snyder, S.; He, C.; Gong, J.; Yue, D.; You, F. *Life-Cycle Analysis of Bioproducts and their Conventional Counterparts in GREET*; Argonne National Lab.: Argonne, IL, USA, 2015.

38. Börjesson, P.; Tufvesson, L.; Lantz, M. *Life Cycle Assessment of Biofuels in Sweden*; Lund University: Lund, Sweden, 2010.

39. El-Suleiman, A.; Anosike, N.; Pilidis, P. A preliminary assessment of the initial compression power requirement in CO2 pipeline Carbon Capture and Storage (CCS) technologies. *Technologies* **2016**, *4*, 15. [CrossRef]

40. *Övervakningsrapport Avseende Skattebefrielse för Flytande Biodrivmedel Under 2018. Report Nr 2019-002678*; Swedish Energy Agency: Estone, Sweden, 2019.

41. European Union Fuel Quality Directive 2009/30/ECof the European Parliament and of the Council of 23 April 2009 Amending Directive 98/70/EC as Regards the Specification of Petrol, Diesel and Gas-Oil and Introducing a Mechanism to Monitor and Reduce Greenhouse Gas Emissions and Amending Council Directive 1999/32/EC as Regards the Specification of Fuel Used by Inland Waterway Vessels and Repealing Directive 93/12/EEC. Available online: https://eur-lex.europa.eu/LexUriServ/LexUriServ.do?uri=OJ:L:2009:140:0088:0113:EN:PDF (accessed on 23 April 2009).

42. European Union Council Directive (EU) 2015/652 of 20 April 2015 Laying Down Calculation Methods and Reporting Requirements Pursuant to Directive 98/70/EC of the European Parliament and of the Council Relating to the Quality of Petrol and Diesel Fuels. Available online: https://eur-lex.europa.eu/legal-content/EN/TXT/PDF/?uri=CELEX:32015L0652&from=EN2015 (accessed on 20 April 2015).

43. Swedish Tax Agency Skattesatser och Växelkurser. The Swedish Tax Agency. Available online: https://www.skatteverket.se/foretagochorganisationer/skatter/punktskatter/energiskatter/skattesatserochvaxelkurser.4.77dbcb041438070e0395e96.html (accessed on 13 November 2019).

44. SPBI Utveckling av Försäljningspris för Bensin, Dieselbränsle Och Etanol. Available online: https://spbi.se/statistik/priser/ (accessed on 4 October 2019).

45. *Scenarier Över Sveriges Energisystem 2018*; Swedish Energy Agency: Estone, Sweden, 2019.

46. Becker, N.; Björnsson, L.; Börjesson, P. *Greenhouse Gas Savings for Swedish Emerging Lignocellulose-Based Biofuels-Using the EU Renewable Energy Directive Calculation Methodology. Report No 104*; Environmental and Energy Systems Studies, Lund Univeristy: Lund, Sweden, 2017.

47. Lantz, M.; Prade, T.; Ahlgren, S.; Björnsson, L. Biogas and ethanol from wheat grain or straw: Is there a trade-off between climate impact, avoidance of iLUC and production cost? *Energies* **2018**, *11*, 2633. [CrossRef]

48. ISO, 14044: 2006. *Environmental Management-Life Cycle Assessment-Requirements and Guidelines*; European Committee for Standardization: Brussels, Belgium, 2006.

49. Camia, A.; Robert, N.; Jonsson, R.; Pilli, R.; García-Condado, S.; López-Lozano, R.; Van der Velde, M.; Ronzon, T.; Gurría, P.; M'barek, R. *Biomass Production, Supply, Uses and Flows in the European Union. First Results from an Integrated Assessment*; Publications Office of the European Union: Luxembourg, 2018.

50. Soam, S.; Kapoor, M.; Kumar, R.; Borjesson, P.; Gupta, R.P.; Tuli, D.K. Global warming potential and energy analysis of second generation ethanol production from rice straw in India. *Appl. Energy* **2016**, *184*, 353–364. [CrossRef]

51. Börjesson, P.; Lundgren, J.; Ahlgren, S.; Nyström, I. *Sustainable Transportation Biofuels Today and in the Future: Summary*; The Swedish Knowledge Centre for Renewable Transportation Fuels: Göteborg, Sweden, 2016.

52. Zhao, L.; Zhang, X.; Xu, J.; Ou, X.; Chang, S.; Wu, M. Techno-economic analysis of bioethanol production from lignocellulosic biomass in China: Dilute-acid pretreatment and enzymatic hydrolysis of corn stover. *Energies* **2015**, *8*, 4096–4117. [CrossRef]

53. Stephen, J.D.; Mabee, W.E.; Saddler, J.N. Will second-generation ethanol be able to compete with first-generation ethanol? Opportunities for cost reduction. *Biofuels Bioprod. Biorefining* **2012**, *6*, 159–176. [CrossRef]

54. Stephen, J.D.; Mabee, W.E.; Saddler, J.N. The ability of cellulosic ethanol to compete for feedstock and investment with other forest bioenergy options. *Ind. Biotechnol.* **2014**, *10*, 115–125. [CrossRef]

55. *Swedish Parliament Fossilfria drivmedel för att minska transportsektorns klimatpåverkan*; Report 2017/18:RFR13; Traffic Committee: Stockholm, Sweden, 2018.